VISTAS IN NANOFABRICATION

edited by Faiz Rahman

VISTAS IN NANOFABRICATION

PAN STANFORD PUBLISHING

Published by

Pan Stanford Publishing Pte. Ltd.
Penthouse Level, Suntec Tower 3
8 Temasek Boulevard
Singapore 038988

Email: editorial@panstanford.com
Web: www.panstanford.com

British Library Cataloguing-in-Publication Data
A catalogue record for this book is available from the British Library.

Vistas in Nanofabrication

ISBN 978-981-4364-56-0 (Hardcover)
ISBN 978-981-4364-57-7 (eBook)

Printed in the USA

Contents

Preface

New materials and devices derived from the application of manufacturing technologies where objects are manipulated at ultra-small scales are becoming gradually commonplace. Modifying materials at micrometre and nanometre scales is often crucial to endowing them with properties that are not found in the base material. Sometimes this is done to gain economic and performance benefits, such as in the manufacture of integrated circuits with ever-smaller features. At other times, nanofabrication is utilised to obtain completely new functionalities, such as in making antireflection structures on plastics and glasses. The relatively new discipline of nanotechnology is now finding increasing use in the manufacture of a wide variety of products, ranging from pharmaceuticals and performance chemicals to apparels and electronic devices. This range continues to expand as new process tools and technologies are developed in research laboratories around the world, on a daily basis. It is now a firmly established fact that the importance of nanotechnology will only increase in the years to come as it makes further inroads into almost every area of human activity. We are already seeing the migration of nanofabrication technologies from the traditional area of electronics manufacturing to other fields such as environmental protection, high-performance sporting goods, manufacture of decorative objects, and other products for everyday use. Despite the increasing use of nanotechnology for producing such goods, its primary application remains in electronics and optoelectronics. This is understandable because further increases in performance of electronic and optoelectronic components will come, to a large extent, from the application of clever nanofabrication techniques in device manufacture. Thus, for example, the continued miniaturization of silicon integrated circuits will become impossible as minimum feature dimensions gradually decrease below 10 nm and then new materials and device architectures — driven by continuing advances in nanotechnology — will be needed to maintain the evolution of circuits and devices towards even higher performance levels. For reasons such as this, a great deal of effort is being invested in developing and identifying new materials and processes that can sustain the industries of the future. The contents of this book provide a glimpse of the work being carried out by nanotechnologists in developing novel technologies for material manipulation and structural nanofabrication. Each chapter presents the recent work of a leading researcher or a research group working at the frontiers of nanotechnology research. A special section at the end of the book presents a collection of micrographs that highlight a variety of structures being created at micron and sub-micron scales. Following is a brief description of the chapters.

The use of nanosphere lithography for patterning dense surface features is described by Hirotaka Oshima from Fujitsu Laboratories in Chapter 1. This is followed by a detailed overview of dry etching technologies for semiconductor manufacture in Chapter 2, by S. J. Pearton from the University of Florida. Basudev Lahiri from the National Institute of Standards and Technology describes his work on split ring-based metamaterials in Chapter 3. I and my colleagues present our work on the fabrication of nanotextured photonic crystal light-emitting diodes in Chapter 4. Chapter 5, contributed by Jin-Seo Noh and colleagues from Yonsei University, deals with the fabrication of nano-wires. A nanotrench-based process for advanced photolithography is described in Chapter 6, by Jean-Francois Dayen and colleagues from the University of Strasbourg. Joong-Mok Park and colleagues from a US research consortium in Iowa describe their work on high-aspect-ratio structures for transparent electrodes in Chapter 7. The fabrication of nano-gap electrodes by novel nanofabrication techniques is described in Chapter 8 by Luis De Los Santos Valladares from an international research collaboration. Chapter 9 contains a description of nanometre-scale processing by tribological techniques by Shojiro Miyake from Nippon Institute of Technology and Mai Wang from the OSG Corporation. Sumita Santra and colleagues from the University of Cambridge and the Indian Institute of Technology describe the integration of nanomaterials in CMOS processing technology in Chapter 10. The next chapter, by Michal Urbanek and Tomáš Šikola from Brno University of Technology describes the use of focussed ion beam techniques for making metallic nanostructures. The section titled "Technology Showcase" presents a compilation of micrographs from the world of nanotechnology.

In putting together this collection of contemporary work on nanofabrication technologies, the editor and the contributors hope to highlight some of the most rapidly developing techniques for micro- and nano-manipulation for modern nanoscale device fabrication. This collection of chapters from leading technologists provides a good survey of the state of the art in some of the most active research areas in nanotechnology at present. We hope that this book will be of use to both new researchers involved with nanofabrication technologies and practicing engineers and scientists who wish to update their knowledge in this fast-changing field.

Faiz Rahman
Glasgow, United Kingdom
June 2012

Chapter 1

Nanosphere Lithography for High-Density Nanopatterning

Hirotaka Oshima
Fujitsu Laboratories Ltd. 10-1 Morinosato-Wakamiya, Atsugi, 243-0197, Japan
oshimah@jp.fujitsu.com

In this chapter, we describe colloidal self-assembly pathways towards fabricating ultrahigh-density ordered nanostructures over large areas. Colloidal self-assembly directed by templates has been shown to enable the formation of controlled nanosphere arrays with periods well below 100 nm. We developed our nanopattern transfer techniques using reactive ion etching and mask layers for fabrication of nanopatterns from dense colloidal arrays and for increasing the pattern height. Their combination with nanoimprint lithography may provide a new technology for high-resolution, high-throughput, and low-cost nanopattern formation.

1.1 Introduction

Photolithography is the most reliable and economical lithography technology for industrial micro- and nanofabrication [1]. This is

Vistas in Nanofabrication
Edited by Faiz Rahman

ISBN 978-981-4364-56-0 (Hardcover), 978-981-4364-57-7 (eBook)
www.panstanford.com

especially true for the semiconductor and magnetic recording industries where the current characteristic feature sizes reach down to sub-100 nm. In these technologies, the fabrication of even smaller nanostructures is often desirable to further improve device properties. Miniaturization technology is also desired in other fields in order to realize new functional devices. In state-of-the-art photolithography systems, excimer laser sources and complex immersion lens systems are used to enhance resist-pattern resolution; they enable pattern sizes smaller than UV laser wavelengths. To fabricate far smaller nanostructures, however, greater progress is required in photolithography technology [2,3]. Extreme UV (EUV) light sources of far shorter wavelengths, their specific optical systems, photoresist materials, and pattern transfer techniques have all been extensively studied for higher resolution photolithography. Nevertheless, their industrial application remains challenging [4]. There are other lithography technologies such as electron beam (EB) lithography, focused ion beam lithography, and scanning probe lithography, which can be used, to some extent, to fabricate smaller nanopatterns than conventional photolithography. Their throughput is, however, very low at the highest resolution due to their serial writing characteristic, while the equipment is rather expensive for large scale use. Direct use of these technologies for manufacturing would critically deteriorate throughput performance and result in a tremendous increase in product costs.

Instead of photoexposure and the development of photoresists, nanoimprint lithography uses moulding with a stamp and etching to create resist patterns [5–7]. This technique is promising since nanoimprint lithography enables ultrahigh-resolution large-area patterning in a single, fast step in thermally or UV-curable resist. There are, nonetheless, still some technological challenges in nanoimprint lithography technology, particularly in nanoimprint mould fabrication for ultrahigh-density nanopatterning. For the mechanical moulding of resist, nanostructures with the same dimensions are the prerequisite for a mould. It means that the original nanostructures of a mould must be fabricated by other high-resolution lithography techniques. Moreover, mould nanopatterns must be of sufficiently large aspect ratio and sharpness to maintain the fidelity of nanopatterns imprinted into the resist during the subsequent etching of residual resist in the compressed areas.

A residual layer of resist at the bottom is essential for uniform nanoimprints without defects; therefore, an etching process for residual resist removal is inevitable. Since nanoimprint moulds can be used repeatedly, stringent cost and time requirements for manufacturing can be alleviated. At this stage, nanoimprint moulds made of Si or quartz having patterns with sub-100 nm pitch and around 30 nm size are commercially available thanks to EB lithography and dry etching techniques. For smaller and more cost-effective nanopatterns, however, further studies are required for nanofabrication technology to improve beyond the resolution limit of current high-resolution lithography and to reduce the time and cost for industrial applications.

For the nanofabrication of periodic micro or nanostructures over macroscopic areas, self-assembly of materials may offer an alternative bottom-up approach [8]. Combined with pattern transfer techniques, it would be useful for producing very large numbers of regular small structures at an affordable cost. Such techniques have been quite powerful tools in fabricating various kinds of nanostructures at the laboratory level. However, controllability of patterns, structures, materials, and manufacturing processes is limited and a long way from most industrial requirements. In addition to the positive properties of nanostructures, commercialization requires manufacturing processes that are reliable, scalable, and cost- and time-effective; it is no easy matter to realize processes that satisfy them all. Progress in materials science and technology, on the other hand, has allowed the emergence of a wide range of new device concepts that require dense periodic arrays of simple but very small nanostructures over large length scales: for instance, magnetic patterned media [9,10], quantum dot devices [11,12], and optical metamaterials [13,14], as will be discussed briefly in the last section. Self-assembly of materials has shown excellent potential by dint of suitable fabrication techniques. In particular, several important advantages have been found in the directed self-assembly of colloidal nanoparticles. They can be uniform, physically and chemically stable, mass-producible, and naturally organized into remarkably ordered periodic structures over macroscopic length scales. Their physical and chemical properties also enable effective subsequent nanopattern transfer [15]. In combination with nanoimprint lithography, therefore, this would resolve issues of controllability and manufacturability and make available various types of nanostructured patterns at low cost and high throughput. In addition, arrays of functional nanoparticles exhibit promising properties for devices such as biotips [16] and chemical sensors [17]. By taking them all into account, we believe that colloidal self-assembly and its applications are worth further research and development from the perspective of manufacturing and commercialization.

In the next section, we introduce colloids and their self-assembly behaviour onto substrates. We then show that the use of templates drastically improves their arrangement and possibly enables precise control of their placement, although the particle order generally deteriorates for smaller nanospheres. In the third section of the chapter, nanopattern transfer techniques called nanosphere lithography are described. In particular, we focus our attention on high-density nanopatterning by nanosphere lithography. In the last section, we discuss future research issues for high-density nanosphere lithography and emerging fields to which these techniques are possibly applicable.

1.2 Colloidal Self-Assembly for Nanopatterning

1.2.1 Colloids

A colloid can be defined as a mixture in which finely divided particles are dispersed throughout another substance. They may be gaseous, liquid, or solid; colloidal particles range in size between nanometres and micrometres. Colloids are found everywhere: cloud, ink, opal, to name a few. Modern chemistry has made it possible to produce monodispersed solid particles in liquid solvents. One of the most important and interesting properties that those colloidal particles generally show is the spontaneous formation of periodic particle arrays called colloidal self-assembly. Under some conditions, particles can be arranged regularly, mostly in the close-packed structure, and they form colloidal crystals [18]. Highly uniform polystyrene (PS) or silica colloidal spheres are typically used for good quality colloidal crystal lattices. Such nanosphere arrays can be utilized as lithographic masks [19], and the techniques are often called "nanosphere lithography" [20].

1.2.2 Formation of Colloidal Crystals

Formation of two-dimensional (2D) arrays of colloidal nanospheres plays an important role in nanosphere lithography. Colloidal crystals can be formed on a substrate from colloidal suspensions by solvent evaporation [21]. In that method self-assembly is induced at the meniscus (Fig. 1.1a). A meniscus is the curve in the liquid surface near a solid wall; it stems from surface tension and wettability [22]. At the meniscus, solvent convective flow supplies colloidal particles to the edge. At the same time, solvent evaporation makes the solvent layer at the meniscus thinner and causes attractive capillary forces between the colloidal particles in the layer. These effects lead to aggregation of colloidal particles on a solid surface [23,24]. When the supply and aggregation of colloidal particles and meniscus displacement by evaporation are all balanced, they result in the 2D growth of colloidal crystals on the substrate. The meniscus displacement can also be controlled by pulling up the substrate immersed in the colloidal suspension – a so-called dip-coating technique [25,26]. By this technique, one can deposit an almost uniform monolayer of colloidal particles onto a large wafer. Self-assembly behaviour at the meniscus depends on the physical and chemical interactions between colloidal particles, particle and solvent, particle and substrate, air and solvent, and solvent and substrate. These various interactions at the meniscus in dip coating self-assembly are schematically depicted in Fig. 1.1b.

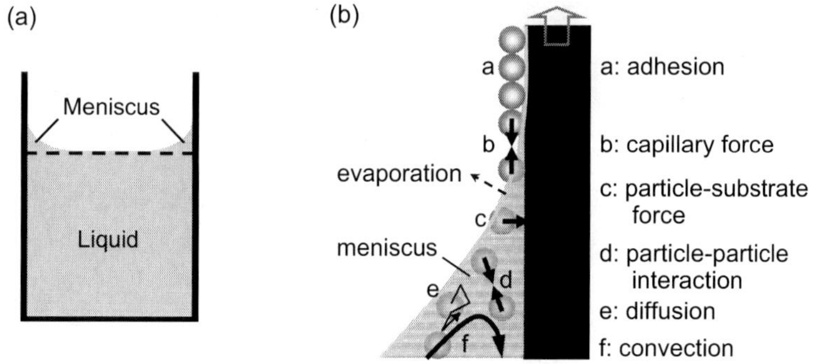

Figure 1.1 Schematic illustrations of (a) a meniscus and (b) interactions at a meniscus in dip-coating process: a, particle adhesion to substrate. b, capillary force between particles. c, particle-substrate force. d, particle-particle interactions in liquid. e, Brownian diffusion. f, movement by convection.

Self-assembled 2D colloidal lattices can also be formed by other techniques. By spin coating the colloidal suspension onto a substrate, solvent evaporation and colloid aggregation can be accelerated. Single-layered structures of colloidal spheres over a macroscopic area can be obtained by using a surfactant that helps to wet the whole substrate surface and helps obtain electrostatic repulsion between the colloidal spheres and the substrate [27]. Surface-modified colloidal spheres can also be trapped at the air-liquid interface. Using the Langmuir-Blodgett coating process, they can be transferred onto a substrate forming a dense monolayer of the colloidal spheres [28]. Application of an alternating electric field drives colloidal microspheres to form 2D ordered structures near the electrode surface in the suspension, which probably stems from electrohydrodynamic flow [29]. Three-dimensional (3D) colloidal lattices would also be necessary for applications such as photonic crystals. A simple method for 3D crystal formation is to use sedimentation in a gravitational field [30]; however, this technique is effective only when colloidal spheres are large enough for sedimentation [31]. Injection of a colloidal suspension into a physically confined cell also induces 3D crystallization of colloidal spheres [32].

1.2.3 Control of Colloidal Crystal Lattices

Colloidal crystal layers deposited onto flat substrates usually contain crystal domains. Domain sizes depend on the materials and methods used for deposition. A crystal domain of tens of micrometres in length with some defects can typically be found in 2D self-assembly of monodispersed PS or silica nanospheres of a few hundreds of nanometres in diameter. An example of such a colloidal crystal is presented in Fig. 1.2. Techniques for producing defect-free perfect colloidal crystals over large areas are, thus, worth investigating. For example, Ho *et al.* used 2D self-assembly of surface-modified PS spheres at the air–water interface and showed defect-free regions larger than $100 \times 100\ \mu m^2$ and $20 \times 20\ \mu m^2$ for 1 μm and 220 nm spheres on substrates, respectively [33]. In addition, they reported that in those experiments, $1 \times 1\ cm^2$ areas were uniformly covered without major cracks or vacancies.

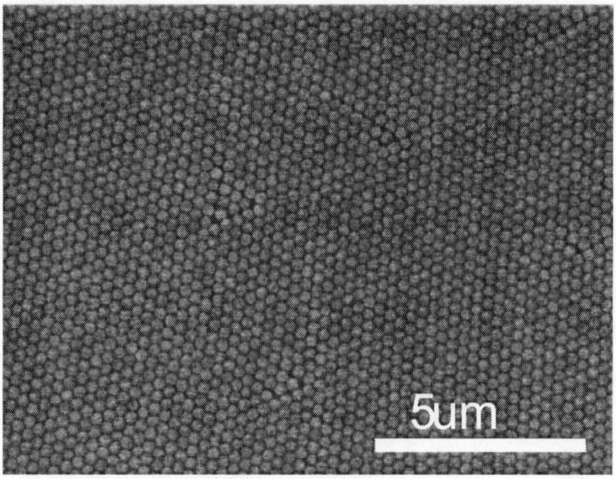

Figure 1.2 SEM image of a silica nanosphere array self-assembled onto a wafer by a dip-coating technique. The diameter of the monodispersed silica nanospheres is 300 nm.

In close packing, periods of sphere arrays are determined by their diameter. To obtain higher areal density patterns, therefore, smaller spheres are necessary as building blocks. Unfortunately, the smaller PS or silica nanospheres that are commercially available show smaller colloidal crystal domain sizes. In dip-coating experiments using colloidal silica well below 100 nm in diameter, for instance, one finds many cracks and defects in the colloidal crystal lattices. It may be due to weakening of the capillary forces [34] or slight deterioration of the nanosphere uniformity in size and shape. In the spin-coating method, Deckman *et al.* have observed that ordering does not effectively occur with colloidal polymer nanospheres smaller than about 50 nm [35]. On the other hand, metal nanospheres of sub-10 nm diameters encapsulated by molecules have been synthesized in the laboratory and show the formation of colloidal crystals: for instance, gold [36], silver [37], cobalt [38], and iron-platinum [39]. However, it is still technically difficult to mass-produce monodispersed uniform nanospheres of these materials. Moreover, their homogeneous monolayer deposition over a large substrate is also quite hard and yet to be realized [40]. Considering these results, for application to high-density nanopatterning, a more sophisticated technique is required to control the domain sizes of colloidal crystals, crystal orientations, and ordering patterns.

1.2.4 Template-Directed Colloidal Self-Assembly

In epitaxial film growth of materials, the growth depends on the crystal orientation of a substrate onto which the materials are deposited. It has been shown that the same is true for colloidal crystal growth. In epitaxy, atoms or molecules are first adsorbed at the energetically favoured sites of a substrate surface, and they influence the subsequent film growth. Similarly, in colloidal self-assembly, colloidal particles can compose a large single crystal with well-defined crystal orientations on a suitably prepatterned substrate. The process works via sedimentation and has been referred to as "colloidal epitaxy" [41]. Such template-based techniques have opened up a new way of controlling the arrangement of self-assembled colloidal crystals [42] and aggregates [43,44]. For example, Ye *et al.* showed that 2D regular array structures of colloidal spheres can be obtained on prepatterned substrates and that the structures obtained depend on the ratio between the sphere diameter and the pattern period [45]. Maury *et al.* presented area-selective monolayer deposition onto polymer-imprinted or chemically patterned templates [46].

These studies on template-directed colloidal self-assembly have, however, mostly focused on colloids whose particle diameters are much larger than 100 nm. It is perhaps, in part, because the techniques would be quite suitable for studies of photonic crystal devices. They require precise control of large-area periodic patterns whose periods are comparable to light wavelengths. Furthermore, templates with micrometre-scale patterns can easily be fabricated by conventional lithography techniques. On the other hand, sub-100 nm particles, such as 40 nm gold nanospheres [47] and 40 nm and 70 nm silica nanospheres [48], have hardly showed long-range order when deposited onto relatively large prepatterns. These results suggest that, for directed self-assembly of sub-100 nm nanospheres, much smaller template structures together with the development of associated experimental techniques are necessary. Indeed, as will be described below, cutting-edge experiments reported by several groups have indicated that by developing prepatterned templates and appropriate deposition techniques, directed self-assembly of nanoparticles can be a promising strategy for high-density nanopatterning.

1.2.5 High-Density Ordered Nanosphere Arrays

Pioneering work on the directed self-assembly of sub-100 nm particles has been performed with templates fabricated by near-field optical lithography [49], EB lithography [50], and EUV lithography [51]: Xia *et al.* showed chain-like closely packed lattices of 50 nm gold colloidal nanospheres arranged in three rows that were formed in 120 nm wide trenches, prepared by near-field optical lithography [49]. Cui *et al.* presented single-particle chains and zigzag-shaped particle chains of 50 nm gold nanospheres in narrow trenches, formed by EB lithography [50]. The nanospheres formed periodic structures in the trenches whose widths were about one and two times the diameter of the nanospheres, respectively. Sub-10 nm particles also showed aggregation into the trenches and holes. The particle configurations, however, appeared to be rather random and did not show significant ordering. The development of EUV interference lithography has enabled 1 × 1 mm areas of groove patterns with sub-100 nm pitch to be defined. Juillerat *et al.* demonstrated that 50 nm silica and gold nanospheres self-assembled onto the templates via a dip-coating technique forming regular single-particle chains in the grooves [51]. In addition, although not precisely lined as 50 nm spheres, even 15 nm gold nanoparticles showed a tendency towards chain formation in grooves with a 50 nm pitch. These results clearly illustrate that directed self-assembly for regular pattern formation is possible even with sub-100 nm particles when the dimensions of the template patterns are comparable to particle diameters.

A much wider variety of applications and fundamental research activities will be possible by precise control of smaller nanoparticle arrays by larger template prepatterns. Recently, we have developed fabrication processes for forming highly ordered 2D hexagonal arrays of 25 nm silica nanospheres on prepatterned substrates [52]. Typical results of the study are presented in Fig. 1.3a,b. The templates in the study had 220 nm pitch line and space patterns over 2 mm wide areas and were fabricated by nanoimprint lithography. Such techniques, therefore, facilitate the fabrication of dense regular nanosphere arrays with a pitch of a few tens of nanometres over large areas. Moreover, the use of inexpensive nanoparticles and low-cost high-throughput lithography tools reduces their costs and extends patterning areas [53]. We prepared the prepatterns by thermal nanoimprint of poly(methyl methacrylate) (PMMA) spin-coated onto a silica layer on the substrate. After the nanoimprint process, the residual resist layer at the bottom of trenches was removed by an oxygen plasma treatment. Dip coating of the colloidal nanospheres was then performed using a homemade dip-coating system in controlled environmental conditions to achieve homogeneous nanosphere deposition over large areas (Fig. 1.4). We used commercial monodispersed silica nanospheres of 100 nm and 25 nm diameters for these experiments. The nanospheres deposited onto the prepatterns were arranged in each trench in two rows (zigzag chains) and in eight rows with 2D hexagonal close packing for 100 nm and 25 nm nanospheres, respectively (Fig. 1.3a,b). Remarkably, the ordering of the nanospheres is significantly improved by the templates and appears to be much better than those deposited onto a flat substrate. This emphasises the critical importance of the templates for the particle interactions in suspensions. We also found that the deposition of a thin Ta layer on the prepatterns did not cause significant changes in the subsequent self-assembly behaviour of the nanospheres. We infer from the results that physical confinement, rather than the difference in wettability, plays the important role in the directed self-assembly process in such systems.

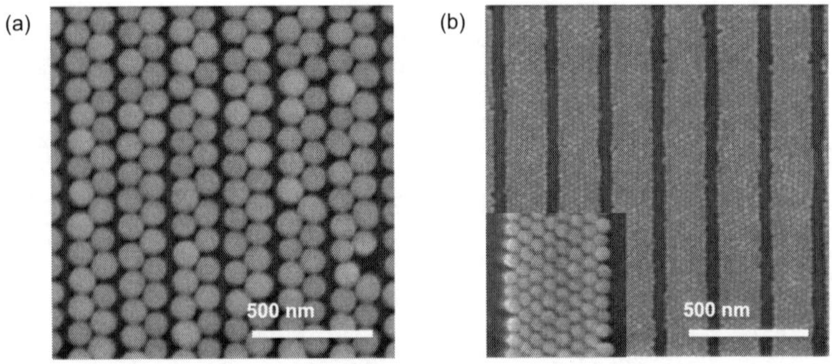

Figure 1.3 Directed self-assembly of silica nanospheres on patterned substrates. SEM images of (a) 100 nm and (b) 25 nm nanosphere arrays. Inset in Fig. 1.3(b) shows eight rows of hexagonally ordered 25 nm nanospheres. Reproduced with permission from Ref. [52]. Copyright 2009, IOP Publishing Ltd.

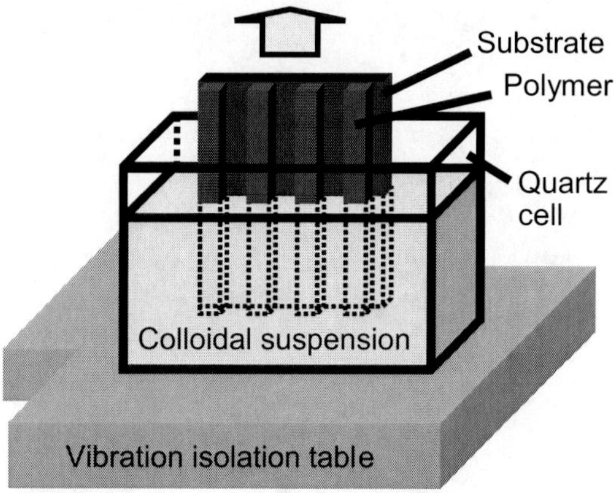

Figure 1.4 Schematic illustration of dip-coating experiment. Reproduced with permission from Ref. [52]. Copyright 2009, IOP Publishing Ltd.

These results strongly suggest that control of sub-10 nm nanoparticle arrays by directed self-assembly will be feasible by developing proper templates and experimental conditions. It will provide well-controlled regular periodic nanopatterns over large areas with much higher areal density than those obtainable by photolithography and even EB lithography [54]. Producing large quantities of sufficiently uniform monodispersed sub-10 nm nanoparticles would be another challenge for the methods to be practicable. Pattern transfer from high-density ordered nanosphere arrays to various materials will open up new approaches to creating nanostructures with new properties and functions. In the next section, we describe nanopattern transfer techniques from colloidal arrays. We pay particular attention to nanoimprint mould fabrication for ultrahigh-resolution lithography technology.

1.3 High-Density Nanosphere Lithography

1.3.1 Colloidal Arrays as Lithographic Masks

Nanosphere lithography is a technique that utilizes colloidal arrays as lithographic masks for nanopatterning of various materials. It is an inexpensive, simple, and parallel nanofabrication technique capable of large-area periodic nanopattern formation. The pattern transfer process from nanosphere arrays can be additive by material deposition or subtractive by etching in the same way as the resist pattern transfer process used with conventional photolithography. Examples are schematically depicted in Fig. 1.5. The arrangement and structure of nanopatterns depend not only on nanosphere arrays but also on the deposition or etching processes.

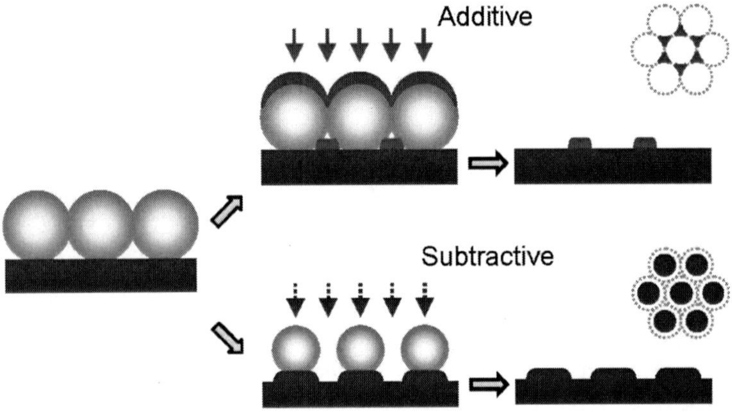

Figure 1.5 Pattern transfer processes in nanosphere lithography: (top) additive pattern transfer by material deposition, (bottom) subtractive pattern transfer by substrate etching. In additive process, nanoparticles are deposited onto a substrate surface through interstices between nanospheres. In subtractive process, nanodots are formed under nanospheres. The diameter of nanospheres can be tuned by an additional etching process before substrate etching.

Monodispersed nanospheres with diameters ranging from micrometres to a few hundred nanometres have been used successfully in nanosphere lithography [15]. Now that considerably higher areal density nanosphere arrays are available, as described in the previous section, it might seem easy to fabricate very small high-density nanostructures by using them for nanosphere lithography. In actuality, however, there remain significant challenges. Nanoscale deposition or etching strongly depends on the dimensions of lithographic patterns; as patterns become smaller, deposition through openings and subsequent mask liftoff become difficult, edge roughness due to etching is enhanced and etching selectivity degrades. Besides, physical and chemical properties of nanospheres may vary depending on their size. Thus, it is not obvious whether straightforward scaling down would be possible in nanosphere lithography. Below we describe some important results and developments of nanosphere lithography for high-density nanopatterning.

1.3.2 Material Deposition

When materials are physically deposited onto a substrate through a nanosphere mask to fabricate ordered nanoparticles, the shape and interparticle spacing of the resultant nanoparticles are determined by the projection of nanosphere mask interstices. In a monolayer of hexagonal closely packed nanospheres, an aperture is made by three adjacent nanospheres, and each nanosphere has six interstices around it (Fig. 1.5). The density of the triangular nanoparticles deposited through the interstices is, therefore, twice as high as that of the mask nanospheres. The distance between the centre of the openings and the nearest edge of the nanospheres is $0.077D$, where D is the nanosphere diameter. Consequently, one can, in principle, fabricate ordered nanoparticle arrays much smaller and of higher density than mask nanospheres [55]. Hulteen and Van Duyne also demonstrated in their pioneering work that by using a double-layer nanosphere mask, hexagonally ordered smaller nanoparticles with a period of D are able to be fabricated [27]. Deposition at a tilted angle also allows the manipulation of the size and shape of resultant nanostructures [56, 57].

Unfortunately, however, when the nanosphere size is reduced, it becomes more and more difficult to deposit nanostructures of sufficient thickness by this technique. We tried to perform physical vapor deposition (PVD) of several materials onto a substrate through 100 nm and 25 nm silica nanosphere arrays: 100 nm nanospheres allowed less than 10 nm deposition, while 25 nm nanospheres did not permit deposition of even a few nm thickness. Further increase in deposition time merely leads to thicker deposition onto nanospheres, which prevents subsequent nanosphere liftoff by sonification. Consequently, nanoparticles possibly deposited through 25 nm nanospheres were too small or too thin to be detected by either scanning electron microscopy or atomic force microscopy.

These results are most probably due to the plugging of the openings by PVD. Ordinarily, this occurs when filling narrow trenches and vias of high aspect ratio by PVD techniques, such as evaporation and sputtering [58]. Lateral deposition onto the upper sidewalls results in a narrowing and eventual closure of the interstices. Since the distance between the centre of the openings and the nearest edge of close-packed 25 nm nanosphere array is only 1.9 nm, it is very likely that only a few nm of deposition significantly reduces or even closes these openings. Such overhang was indeed a major problem in metallization technology in the integrated circuit industry [59], until new damascene electroplating techniques were developed. For additive pattern transfer from high-density nanosphere arrays, the development of a new process utilizing chemical deposition techniques is likewise necessary.

1.3.3 Substrate Etching

Periodic nanodot arrays can also be fabricated in a subtractive manner by etching a substrate surface through a nanosphere mask. Anisotropic selective reactive ion etching (RIE) techniques are very useful for such pattern transfers, and nanodot or nanotip arrays of various materials have been fabricated to date [15]. Typical experiments are performed using submicron PS nanospheres on a silicon wafer or a glass substrate [60]. In such experiments, the diameters of ordered PS nanospheres are first tuned by O_2 RIE; substrate surfaces are subsequently etched using, for example, CF_4 gas. Cheung *et al.* employed the Bosch process, which uses alternate etching and passivation processes, for deep etching of silicon under nanospheres. They fabricated silicon nanopillars with an aspect ratio up to 10 using 500 nm PS nanosphere arrays as a mask layer [61]. However, we found that it becomes difficult to obtain well-ordered arrays of PS nanospheres of sub-100 nm diameter even on a template. Moreover, PS nanospheres smaller than 100 nm are etched more easily by RIE. They are etched even faster than substrate materials during substrate etching and their shape drastically deteriorates, probably because of unavoidable physical etching in RIE.

In contrast, we found that 100 nm silica nanospheres show good ordering (Fig. 1.3a) as well as remarkable pattern transfer characteristics. We have demonstrated that by RIE with CF_4 gas ordered 100 nm silica nanosphere arrays can be directly transferred into arrays of hemi-ellipsoidal nanodots on silica substrate surfaces [52]. This is shown in Fig. 1.6. Here, owing to similar etching rates for the 100 nm nanospheres and the surface layer, the nanodot height reaches about 100 nm when the nanospheres vanish. Fabrication of firm silica nanodot arrays is of significant technological importance: they have stable physical and chemical properties, are UV transparent, and thus can be used as a thermal as well as a UV nanoimprint mould. As shown in Fig. 1.6c, we have demonstrated that nanostructures fabricated on the substrate can be inversely transferred to a polymer resist film through UV nanoimprinting by using nanodot structures as a mould [52]. Notice that the UV nanoimprint experiment can be performed repeatedly.

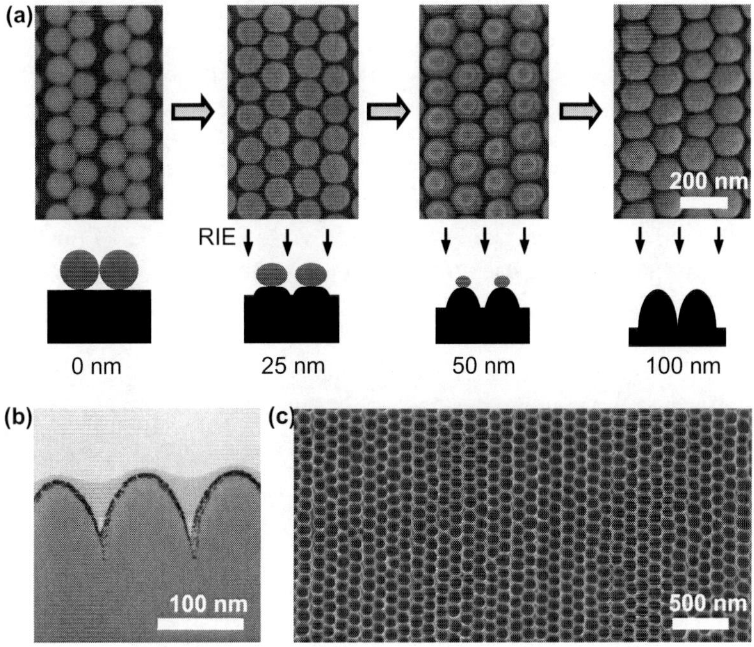

Figure 1.6 Direct pattern transfer from 100 nm nanosphere arrays. (a) SEM images taken after CF_4 RIE with nominal etching depths of 0, 25, 50, and 100 nm, and corresponding schematic representations. (b) Transmission electron micrograph of the nanodots. The sample surface was coated with a Pt layer for SEM observation. (c) SEM image of the polymer surface after UV nanoimprinting. Reproduced with permission from Ref. [52]. Copyright 2009, IOP Publishing Ltd.

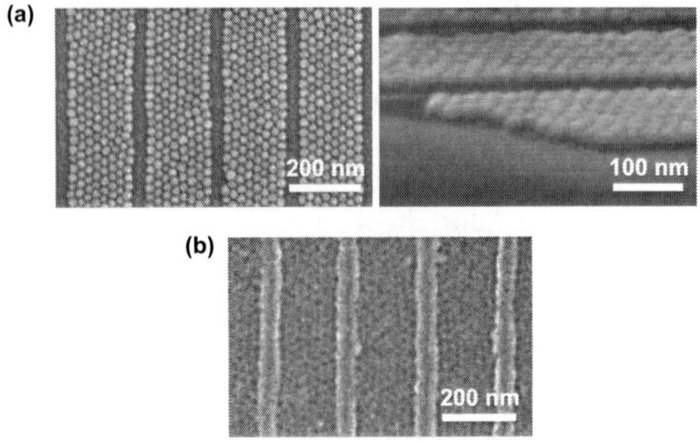

Figure 1.7 (a) Top view and tilted view SEM images of the silica nanodot arrays fabricated by the direct pattern transfer. (b) SEM image of the polymer surfaces after UV nanoimprint using the nanodot arrays shown in (a) as the nanoimprint mould. Reproduced with permission from Ref. [52]. Copyright 2009, IOP Publishing Ltd.

More importantly, for high-density nanopatterning, the same direct pattern transfer technique using CF_4 RIE can also be applied to arrays of 25 nm silica nanospheres shown in Fig. 1.3b. After etching the sample with CF_4 gas until the 25 nm nanospheres disappeared, periodic nanodot arrays with 25 nm pitch were formed at the silica layer (Fig. 1.7a). Here the nanodots are rounded with an average height of 16 nm. The decrease in height indicates that 25 nm silica nanospheres are etched faster than the silica substrate layer itself, in contrast to 100 nm diameter nanospheres. The 25 nm pitch nanodot arrays were also used as a UV nanoimprint mould. However, because of the decrease in height and sharpness, the hexagonal patterns formed on a polymer surface are rather indistinct (Fig. 1.7b). Further pattern transfer from the polymer nanopatterns to other materials would thus be difficult to achieve. These results suggest that one can realize a higher aspect ratio of nanodots for well-defined nanoimprint lithography. Development of a lithographic method for that purpose will be described next.

1.3.4 High-Density Nanopattern Transfer

In order to fabricate high-aspect-ratio, high-density, large-area nanodots by RIE, etching should be sufficiently selective, anisotropic, and uniform. We should note that even the aforementioned Bosch process for deep-RIE is not suitable for that purpose because of its inevitable lateral etching (scalloping) of silicon with SF_6 gas during nanostructure formation. Aiming at satisfying all the conditions required, we have developed a lithographic method that uses etch mask layers inserted below nanospheres [52]. In that method, we utilize a thin Ru layer sputter-deposited onto the silica surface of a substrate for a mask layer. A SiO_x layer is then sputter-deposited onto the Ru layer to obtain the same surface chemistry for the self-assembly of 25 nm silica nanospheres on it. This arrangement works because ruthenium and silica can be good etch masks for each other by using O_2 gas for Ru etching and CF_4 gas for silica etching.

The pattern transfer process consists of three steps (Fig. 1.8a): (1) from silica nanospheres to SiO_x layer by CF_4 RIE, (2) from SiO_x layer to Ru layer by O_2 RIE, and (3) from Ru layer to silica surface by CF_4 RIE. In these steps, silica is etched by CF_4 RIE about ten times faster than a Ru layer, and a Ru layer can be effectively etched by O_2 RIE while silica is hardly etched with O_2 gas. We should also point out that all the materials are generally used in sputter deposition and RIE technologies and are quite manageable, which is of particular importance for their industrial applications.

In Fig. 1.8b, the pillar-shaped nanodot (nanotip) arrays fabricated by the above mentioned mask-layer method are presented. The average height of the nanodots is estimated to be 45 nm, and their average bottom diameter is 19 nm with a standard deviation of 2 nm. An aspect ratio of more than two is thus achieved. A polymer surface, UV-nanoimprinted by the high-aspect-ratio nanodot arrays is presented in Fig. 1.8c. The hexagonal patterns are quite clear, reflecting the high aspect ratio of the nanodots.

1.4 Future Prospects and Applications

1.4.1 Challenges and Future Directions

The fabrication of nanoimprint moulds using high-density nanosphere lithography is, of course, only a beginning. There are a variety of research issues related to industrial applications, manufacturing, and commercialization. We provide some examples below.

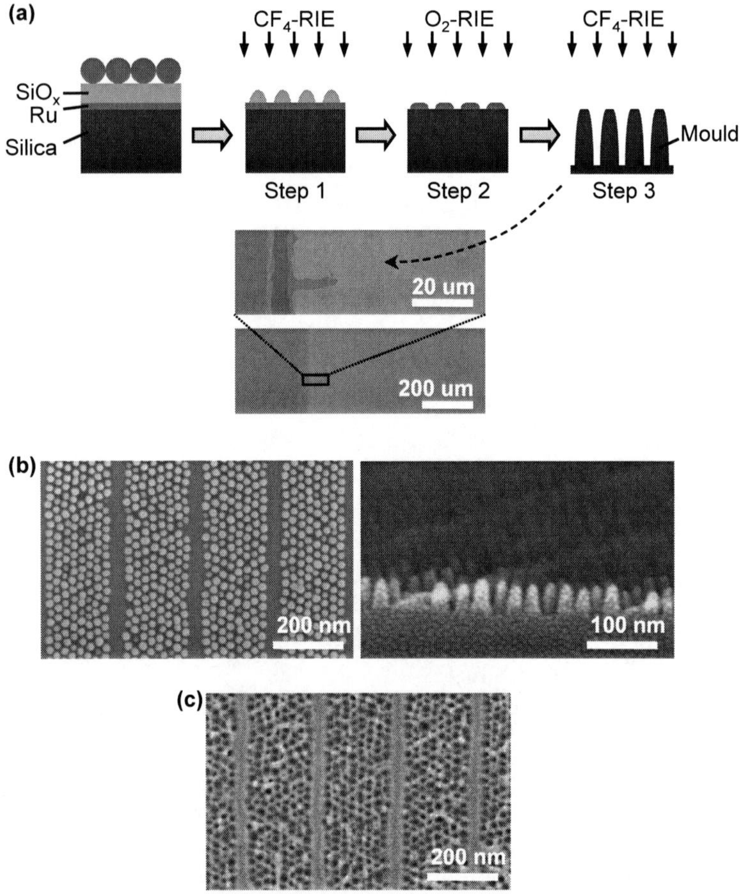

Figure 1.8 (a) Schematic illustrations of the mask-layer method and large-area SEM images of the nanopatterns. First, the silica nanosphere arrays are arranged on the Ru/SiO_x layers. Then, the patterns are transferred from nanospheres to SiO_x layer by CF_4 RIE (step 1), from SiO_x layer to Ru layer by O_2 RIE (step 2), and from Ru layer to silica layer by CF_4 RIE (step 3). The nanopatterns are fabricated uniformly in the bright regions of the large-area images. (b) Top view and tilted view SEM images of the pillar-shaped nanodots fabricated by the mask-layer method. (c) SEM images of the polymer surfaces after UV nanoimprint using the nanodot arrays shown in (b) as the nanoimprint mould. Reproduced with permission from Ref. [52]. Copyright 2009, IOP Publishing Ltd.

For template-directed colloidal self-assembly to fabricate higher density nanopatterns, large-scale preparation of uniform monodispersed colloidal nanospheres of sub-10 nm diameter is indispensable [62]. Design for functional template patterns and self-assembly technologies for them should both be studied. For example, instead of close-packed structures, nanospheres are able to form separated arrays at suitable pattern edges by using capillary immersion forces [63]. Binary nanoparticles of sub-10 nm diameters show a large variety of excellent binary superlattices [64], which would be quite useful for control of complex nanopatterns. Novel self-assembly techniques may be possible by applying external fields during nanosphere deposition onto templates that are polarized by fields and can interact with nanospheres.

Nanopattern transfer technology is an important research area for advanced lithography. In order to fabricate higher density arrays of higher aspect ratio nanostructures, we would need to delve deeper into nanomaterials science for understanding the behaviour of colloids, etching masks, etching gases, and substrate surfaces. Material compatibility is extremely important for nanopattern transfer. For example, although being sparse and random, Lewis *et al.* fabricated high-aspect-ratio silicon nanopillars by $SiCl_4$ RIE using 15 nm and 10 nm gold colloidal nanoparticle masks [65]. They reported that for deep silicon etching with gold nanoparticle masks $SiCl_4$ gas was better suited than CF_4 gas. Further investigations into nanoimprint lithography would certainly continue. The high resolutions possible with this technique have already been demonstrated in many ways [66]. In devices, however, various patterns and materials are used, which would give rise to diverse technological problems. Reduction of size distributions, defects, and edge roughness in nanofabrication will be especially important [67].

1.4.2 Possible Applications

As mentioned earlier, high-density nanosphere lithography will be applicable to a variety of emerging technologies, particularly in combination with nanoimprint lithography. These are, for instance, quite useful for the fabrication of nanomagnetic devices, including magnetic patterned media. Future ultrahigh-density magnetic recording media will require patterned magnetic nanodot arrays for information bits instead of conventional granular magnetic thin films [3]. Recording densities greater than 1 Tbit/in^2 require full-pitch lithographic resolutions better than 27 nm and thus will rely on highly advanced nanopatterning technologies.

We have studied the fabrication of nanomagnetic arrays for magnetic media using nanoimprint lithography. In Fig. 1.9a, we show ordered anodic alumina nanohole arrays and cobalt nanopillars electrodeposited in the nanoholes. Here the regular nanohole positions were defined by a nanoimprint lithography technique that we have developed [10]. We also found that magnetic nanodot arrays can be fabricated simply by exfoliating magnetic thin films deposited onto nanoimprinted polymer surfaces (Fig. 1.9b); the technique needs neither etching nor liftoff for nanodot formation [68]. Mould patterns define magnetic nanodots in this technique. Nanosphere lithography for nanoimprint moulds will help increase their density and control their shapes and patterns.

Figure 1.9 (a) SEM images of (top) ideally ordered anodic alumina nanohole arrays fabricated using nanoimprint lithography and (bottom) nanohole patterned media using alumina nanohole arrays, in which cobalt nanopillars are embedded by electrodeposition. Reproduced with permission from Ref. [10]. Copyright 2007, American Institute of physics. (b) (top) SEM image of the exfoliated Co/Pd multilayer nanodot array, (bottom) MFM image of the exfoliated Co/Pd multilayer nanodot array. The MFM image was acquired in the remanent state after first applying a negative saturation field and then exposed to a 1 kOe reversal field. Reproduced with permission from Ref. [68]. Copyright 2008, The Japan Society of Applied Physics.

Magnetic multilayer nanodot arrays can also be used for spintronic devices, such as spin torque nano-oscillators that can generate microwaves by a perpendicular dc current [69]. To obtain spin torque transfer effectively in magnetic junctions, the device element size should be smaller than several tens of nanometres. Increase in density of the element arrays would enhance the level of emission power.

High-density ordered nanodot arrays would also be useful for photonic devices that utilize quantum dots [11,12]. Next generation photovoltaic solar cells may involve quantum dots to enhance their conversion efficiency [12,70]. The dots should be smaller than 10 nm for quantum confinement and be dense enough for the overlap of wavefunctions to form new band structures to reduce the various losses in a solar cell. Uniformity and regularity of the quantum dot arrays would strongly affect the device performance and its controllability. Ordered quantum dot arrays may also improve device performance of quantum dot lasers [11].

Optical metamaterials is another possible future application of high-density ordered nanostructures [14]. Arrays of artificial structures, such as split-ring resonators, nanorods, and nanofishnets, have enabled negative permittivity and permeability at their specific resonance frequencies (typically microwaves), leading to the observation of negative refraction effects. Such negative index metamaterials would enable super-lenses for imaging beyond the diffraction limit and cloaking devices that make objects invisible. Frequencies for the existence of negative refraction are chiefly determined by the dimensions of the artificial structure arrays. To realize optical metamaterials, their sizes and periods should be much smaller than optical wavelengths [14]. It appears that a combination of nanosphere lithography and nanoimprint lithography would facilitate large-area fabrication of engineered nanostructure arrays for negative index metamaterials for optical applications.

Acknowledgements

We gratefully acknowledge contributions from H. Tamura, M. Takeuchi, K. Itoh, H. Nakao, H. Kikuchi, T. Sato, S. Umehara, A. Inomata, Y. Yanagida, N. Matsushita, H. Komoriya, T. Uzumaki, A. Tanaka, K. Nishio, and H. Masuda, and support from Nippon Shokubai Co. Ltd. and JGC Catalysts and Chemicals Ltd. The author's research was in part financially supported by The New Energy and Industrial Technology Development Organization (NEDO) and Japan Science and Technology Agency (JST).

References

1. Rahman, F. (2008). *Nanostructures in Electronics and Photonics*, ed. Rahman, F., Chapter 1 "From Microstructures to Nanostructures" (Pan Stanford Publishing, Singapore) pp. 1–27.
2. Pease, F., and Chou, S. Y. (2008). Lithography and other patterning techniques for future electronics, *Proc. IEEE*, **96**, pp. 248–270.
3. Terris, B. D., and Thomson, T. (2005). Nanofabricated and self-assembled magnetic structures as data storage media, *J. Phys. D: Appl. Phys.*, **38**, pp. R199–R222.
4. Solak, H. H. (2006). Nanolithography with coherent extreme ultraviolet light, *J. Phys. D: Appl. Phys.*, **39**, pp. R171–R188.
5. Chou, S. Y., Krauss, P. R., and Renstrom P. J. (1996). Imprint lithography with 25-nanometer resolution, *Science*, **272**, pp. 85–87.
6. Haisma, J., Verheijen, M., van den Heuvel, K., and van den Berg, J. (1996). Mold-assisted nanolithography: A process for reliable pattern replication, *J. Vac. Sci. Technol. B*, **14**, pp. 4124–4128.
7. Schift, H. (2008). Nanoimprint lithography: An old story in modern times? A review, *J. Vac. Sci. Technol. B*, **26**, pp. 458–480.
8. Whitesides, G. M., and Grzybowski, B. (2002). Self-assembly at all scales, *Science*, **295**, pp. 2418–2421.
9. Darling, S. B., and Bader, S. D. (2005). A materials chemistry perspective on nanomagnetism, *J. Mater. Chem.*, **15**, pp. 4189–4195.
10. Oshima, H., Kikuchi, H., Nakao, H., Itoh, K., Kamimura, T., Morikawa, T., Matsumoto, K., Umada, T., Tamura, H., Nishio, K., and Masuda, H. (2007). Detecting dynamic signals of ideally ordered nanohole patterned disk media fabricated using nanoimprint lithography, *Appl. Phys. Lett.*, **91**, 022508.
11. Arakawa, Y., and Sakaki, H. (1982). Multidimensional quantum well laser and temperature dependence of its threshold current, *Appl. Phys. Lett.*, **40**, pp. 939–941.
12. Conibeer, G., Green, M., Corkish, R., Cho, Y., Cho, E.-C., Jiang, C.-W., Fangsuwannarak, T., Pink, E., Huang, Y., Puzzer, T., Trupke, T., Richards, B., Shalav, A., and Lin, K.-L. (2006). Silicon nanostructures for third

generation photovoltaic solar cells, *Thin Solid Films*, **511–512**, pp. 654–662.

13. Galisteo, J. F., García-Santamaría, F ., Golmayo, D., Juárez, B. H., López, C., and Palacios, E. (2005). Self-assembly approach to optical metamaterials, *J. Opt. A: Pure Appl. Opt.*, **7**, pp. S244–S254.
14. Shalaev, V. M. (2007). Optical negative-index metamaterials, *Nat. Photon.*, **1**, pp. 41–48.
15. Zhang, J., Li, Y., Zhang, X., and Yang, B. (2010). Colloidal self-assembly meets nanofabrication: From two-dimensional colloidal crystals to nanostructure arrays, *Adv. Mater.*, **22**, pp. 4249–4269.
16. Freeman, R. G., Grabar, K. C., Allison, K. J., Bright, R. M., Davis, J. A., Guthrie, A. P., Hommer, M. B., Jackson, M. A., Smith, P. C., Walter D. G., and Natan, M. J. self-assembled metal colloid monolayers: An approach to SERS substrates, *Science*, **267**, pp. 1629–1632.
17. Holtz, J. H., and Asher, S. A. (1997). Polymerized colloidal crystal hydrogel films as intelligent chemical sensing materials, *Nature*, **389**, pp. 829–832.
18. Xia, Y., Gates, B., Yin,Y., and Yu Lu, Yu. (2000). Monodispersed colloidal spheres: Old materials with new applications, *Adv. Mater.*, **12**, pp. 693–713.
19. Deckman, H. W., and Dunsmuir, J. H. (1982). Natural lithography, *Appl. Phys. Lett.*, **41**, pp. 377–379.
20. Haynes, C. L., and Van Duyne, R. P. (2001). Nanosphere lithography: A versatile nanofabrication tool for studies of size-dependent nanoparticle optics, *J. Phys. Chem. B*, **105**, pp. 5599–5611.
21. Nagayama, K. (1996). Two-dimensional self-assembly of colloids in thin liquid films, *Colloids Surf. A*, **109**, pp. 363–374.
22. de Gennes, P. G., Brochard-Wyart, F., and Quéré, D. (2004). *Capillarity and Wetting Phenomena: Drops, Bubbles, Pearls, Waves* (Springer, New York).
23. Kralchevsky, P. A., and Denkov, N. D. (2001). Capillary forces and structuring in layers of colloid particles, *Curr. Opin. Colloid Interface Sci.*, **6**, pp. 383–401.
24. Rivera, T. P., Lecarme, O., Hartmann, J., Rossitto, E., Berton, K., and Peyrade, D. (2008). Assisted convective-capillary force assembly of gold colloids in a microfluidic cell: Plasmonic properties of deterministic nanostructures, *J. Vac. Sci. Technol. B*, **26**, pp. 2513–2519.

25. Dimitrov, A. S., and Nagayama, K. (1996). Continuous convective assembling of fine particles into two-dimensional arrays on solid surfaces, *Langmuir*, **12**, pp. 1303–1311.

26. Jiang, P., Bertone, J. F., Hwang, K. S., and Colvin, V. L. (1999). Single-crystal colloidal multilayers of controlled thickness, *Chem. Mater.*, **11**, pp. 2132–2140.

27. Hulteen, J. C., and Van Duyne, R. P. (1995). Nanosphere lithography: A materials general fabrication process for periodic particle array surfaces, *J. Vac. Sci. Technol. A*, **13**, pp. 1553–1558.

28. Deckman, H. W., Dunsmuir, J. H., and Gruner, S. M. (1989). Microfabrication cellular phosphors, *J. Vac. Sci. Technol. B*, **7**, pp. 1832–1835.

29. Xie, R., and Liu, X.-Y. (2008). Electrically directed on-chip reversible patterning of two-dimensional tunable colloidal structures, *Adv. Funct. Mater.*, **18**, pp. 802–809.

30. Pysey, P. N., and van Megen, W. (1986). Phase behaviour of concentrated suspensions of nearly hard colloidal spheres, *Nature*, **320**, pp. 340–342.

31. Holgado, M., García-Santamaría, F., Blanco, A., Ibisate, M., Cintas, A., Míguez, H., Serna, C. J., Molpeceres, C., Requena, J., Mifsud, A., Meseguer, F., and López, C. (1999). Electrophoretic deposition to control artificial opal growth, *Langmuir*, **15**, pp. 4701–4704.

32. Park, S. H., and Xia, Y. (1999). Assembly of mesoscale particles over large areas and its application in fabricating tunable optical filters, *Langmuir*, **15**, pp. 266–273.

33. Ho, C.-C., Hsieh, T.-W., Kung, H.-H., Juan, W.-T., Lin, K.-H., and Lee, W.-L. (2010). Reduced saturation magnetization in cobalt antidot thin films prepared by polyethylene oxide-assisted self-assembly of polystyrene nanospheres, *Appl. Phys. Lett.*, **96**, 122504.

34. Kralchevsky, P. A., and Nagayama, K. (2000). Capillary interactions between particles bound to interfaces, liquid films and biomembranes, *Adv. Colloid Interface Sci.*, **85**, pp. 145–192.

35. Deckman, H. W., Dunsmuir, J. H., Garoff, S., McHenry, J. A., and Peiffer, D. G. (1988). Macromolecular self-organized assemblies, *J. Vac. Sci. Technol. B*, **6**, pp. 333–336.

36. Andres, R. P., Bielefeld, J. D., Henderson, J. I., Janes, D. B., Kolagunta, V. R., Kubiak, C. P., Mahoney, W. J., and Osifchin, R. G. (1996). Self-assembly of a two-dimensional superlattice of molecularly linked metal clusters, *Science*, **273**, pp. 1690–1693.

37. Harfenist, S. A., Wang, Z. L., Alvarez, M. M., Vezmar, I., and Whetten, R. L. (1996). Highly oriented molecular Ag nanocrystal arrays, *J. Phys. Chem.*, **100**, pp. 13904–13910.

38. Sun, S., and Murray, C. B. (1999). Synthesis of monodisperse cobalt nanocrystals and their assembly into magnetic superlattices, *J. Appl. Phys.*, **85**, pp. 4325–4330.

39. Sun, S., Murray, C. B., Weller, D., Folks, L., and Moser, A. (2000). Monodisperse FePt nanoparticles and ferromagnetic FePt nanocrystal superlattices, *Science*, **287**, pp. 1989–1992.

40. Kodama, H., Momose, S., Ihara, N., Uzumaki, T., and Tanaka, A. (2003). Disk substrate deposition techniques for monodisperse chemically synthesized FePt nanoparticle media, *Appl. Phys. Lett.*, **83**, pp. 5253–5255.

41. van Blaaderen, A., Ruel, R., and Wiltzius, P. (1997). Template-directed colloidal crystallization, *Nature,* **385**, pp. 321–324.

42. Hoogenboom, J. P., Rétif, C., de Bres, E., van de Boer, M., van Langen-Suurling, A. K., Romijn, J., and van Blaaderen, A. (2004). Template-induced growth of close-packed and non-close-packed colloidal crystals during solvent evaporation, *Nano Lett.*, **4**, pp. 205–208.

43. Yin, Y., Lu, Y., Gates, B., and Xia, Y. (2001). Template-assisted self-assembly: A practical route to complex aggregates of monodispersed colloids with well-defined sizes, shapes, and structures, *J. Am. Chem. Soc.*, **123**, pp. 8718–8729.

44. Wang, D., and Möhwald, H. (2004). Template-directed colloidal self-assembly – the route to 'top-down' nanochemical engineering, *J. Mater. Chem.*, **14**, pp. 459–468.

45. Ye, Y.-H., Badilescu, S., Truong, V.-V., Rochon, P., and Natansohn, A. (2001). Self-assembly of colloidal spheres on patterned substrates, *Appl. Phys. Lett.*, **79**, pp. 872–874.

46. Maury, P., Escalante, M., Reinhoudt, D. N., and Huskens, J. (2005). Directed assembly of nanoparticles onto polymer-imprinted or chemically patterned templates fabricated by nanoimprint lithography, *Adv. Mater.*, **17**, pp. 2718–2723.

47. Fan, F., and Stebe, K. J. (2004). Assembly of colloidal particles by evaporation on surfaces with patterned hydrophobicity, *Langmuir*, **20**, pp. 3062–3067.

48. Huwiler, C., Halter, M., Rezwan, K., Falconnet, D., Textor, M., and Vörös, J. (2005). Self-assembly of functionalized spherical nanoparticles on chemically patterned microstructures, *Nanotechnology*, **16,** pp. 3045–3052.

49. Xia, Y., Yin, Y., Lu, Y., and McLellan, J. (2003). Template-assisted self-assembly of spherical colloids into complex and controllable structures, *Adv. Funct. Mater.*, **13**, pp. 907–918.

50. Cui, Y., Björk, M. T., Liddle, J. A., Sönnichsen, C., Boussert, B., and Alivisatos, A. P. (2004). Integration of colloidal nanocrystals into lithographically patterned devices, *Nano Lett.*, **4**, pp. 1093–1098.

51. Juillerat, F., Solak, H. H., Bowen P., and Hofmann, H. (2005). Fabrication of large-area ordered arrays of nanoparticles on patterned substrates, *Nanotechnology*, **16**, pp. 1311–1316.

52. Oshima, H., Tamura, H., Takeuchi, M., Inomata, A., Yanagida, Y., Matsushita, N., Komoriya, H., Uzumaki, T., and Tanaka, A. Nanopattern transfer from high-density self-assembled nanosphere arrays on prepatterned substrates, *Nanotechnology*, **20**, 455303.

53. Velev, O. D., and Gupta, S. (2009). Materials fabricated by micro- and nanoparticle assembly – the challenging path from science to engineering, *Adv. Mater.*, **21**, pp. 1897–1905.

54. Hosaka, S., Mohamad, Z., Shira, M., Sano, H., Yin, Y., Miyachi, A., and Sone, H. (2008). Extremely small proximity effect in 30 keV electron beam drawing with thin calixarene resist for 20×20 nm^2 pitch dot arrays, *Appl. Phys. Express*, **1**, 027003.

55. Hulteen, J. C., Treichel, D. A., Smith, M. T., Duval, M. L., Jensen, T. R., and Van Duyne, R. P. (1999). Nanosphere lithography: Size-tunable silver nanoparticle and surface cluster arrays, *J. Phys. Chem. B*, **103**, pp. 3854–3863.

56. Haynes, C. L., McFarland, A. D., Smith, M. T., Hulteen, J. C., and Van Duyne, R. P. (2002). Angle-resolved nanosphere lithography: manipulation of nanoparticle size, shape, and interparticle spacing, *J. Phys. Chem. B*, **106**, pp. 1898–1902.

57. Zhang, G., Wang, D., and Möhwald, H. (2007). Ordered binary arrays of Au nanoparticles derived from colloidal lithography, *Nano Lett.*, 7, pp. 127–132.

58. Monteiro, O. R. (1999). Novel metallization technique for filling 100-nm-wide trenches and vias with very high aspect ratio, *J. Vac. Sci. Technol. B*, **17**, pp. 1094–1097.

59. Rossnagel, S. M. (1998). Directional and ionized physical vapor deposition for microelectronics applications, *J. Vac. Sci. Technol. B*, **16**, pp. 2585–2608.

60. Wang B., Zhao W., Chen A., and Chua S.-J. (2006). Formation of nanoimprinting mould through use of nanosphere Lithography, *J. Cryst. Growth*, **288**, pp. 200–204.

61. Cheung, C. L., Nikolić, R. J., Reinhardt, C. E., and Wang, T. F. (2006). Fabrication of nanopillars by nanosphere lithography, *Nanotechnology*, **17,** pp. 1339–1343.

62. Park, J., An, K., Hwang, Y., Park, J.-G., Noh, H.-J., Kim, J.-Y., Park, J.-H., Hwang, N.-M., and Hyeon, T. (2004). Ultra-large-scale syntheses of monodisperse nanocrystals, *Nat. Mater.*, **3**, pp. 891–895.

63. Gordon, M. J., and Peyrade, D. (2006). Separation of colloidal nanoparticles using capillary immersion forces, *Appl. Phys. Lett.*, **89**, 053112.

64. Shevchenko, E. V., Talapin, D. V., Murray, C. B., and O'Brien, S. (2006). Structural characterization of self-assembled multifunctional binary nanoparticle superlattices, *J. Am. Chem. Soc.*, **128**, pp. 3620–3637.

65. Lewis, P. A., Ahmed, H., and Sato, T. (1998). Silicon nanopillars formed with gold colloidal particle masking, *J. Vac. Sci. Technol. B*, **16**, pp. 2938–2941.

66. Austin, M. D., Ge, H., Wu, W., Li, M., Yu, Z., Wasserman, D., Lyon, S. A., and Stephen Y. Chou, S. Y. (2004). Fabrication of 5 nm linewidth and 14 nm pitch features by nanoimprint lithography, *Appl. Phys. Lett.*, **84**, pp. 5299–5301.

67. Chou, S. Y., and Xia, Q. (2008). Improved nanofabrication through guided transient liquefaction, *Nat. Nanotech.*, **3**, pp. 295–300.

68. Oshima, H., Kikuchi, H., Nakao, H., Itoh, K., Morikawa, T., Tamura, H., Nishio,K., and Masuda, H. (2008). Magnetic nanodot array formation by exfoliation of sputter-deposited magnetic thin films, *Appl. Phys. Express*, **1**, 054001.

69. Kaka, S., Pufall, M. R., Rippard, W. H., Silva, T. J., Russek, S. E., and Katine, J. A. (2005). Mutual phase-locking of microwave spin torque nano-oscillators, *Nature*, **437**, pp. 389–392.

70. Tsakalakos, L. (2008). Nanostructures for photovoltaics, *Mater. Sci. Eng. R*, **62**, pp. 175–189.

Chapter 2

Dry Etching of Semiconductors at the Nano- and Micro-Scale

S. J. Pearton
Department of Materials Science and Engineering, PO Box 116400, University of Florida, Gainesville, FL 32611 USA
spear@mse.ufl.edu

In this chapter, wet and dry etching of semiconductors used in optoelectronics is reviewed. In cases where the active region width is large (>3 microns), wet etching provides a simple, effective method for pattern transfer. For more demanding structures and for a greater degree of control during device fabrication, plasma-etching methods offer advantages.

2.1 Introduction

There is generally a need to pattern compound semiconductors when fabricating devices such as ridge waveguide laser diodes, light-emitting diodes, photodetectors, and associated electronics. The patterning is carried out by etching the semiconductor, using dielectric or photoresist masks to protect the active areas. Etch processes may be classified by their rate, selectivity, uniformity, directionality

Vistas in Nanofabrication
Edited by Faiz Rahman

ISBN 978-981-4364-56-0 (Hardcover), 978-981-4364-57-7 (eBook)
www.panstanford.com

(isotropy or anisotropy), surface quality, and reproducibility. All etching processes involve three basic events: (1) movement of the etching species to the surface to be etched, (2) chemical reaction to form a compound that is soluble in the surrounding medium, and (3) movement of the by-products away from the etched region, allowing fresh etchant to reach the surface. Both (1) and (3) usually are referred to as diffusion, although convection may be present. The slowest of these processes primarily determines the etch rate, which may be diffusion or chemical-reaction limited. There are two different etching methods by using two quite different media: liquid chemicals (wet etching) and reactive gas plasmas (dry etching). Wet etching is performed by immersing the wafers in an appropriate solution or by spraying the wafer with the etchant solution. Wet-chemical etching is superior to dry etching in terms of effectiveness, simplicity, low cost, low damage to the wafer, high selectivity, and high throughput [1]. However, the main limitations of wet etching include its isotropic nature which results in roughly equal rate of removal of material in all directions, making it incapable of patterning sub-micron features, and the need for the disposal of large amounts of corrosive and toxic materials. Dry etching methods became favorable etching processes for integrated circuit manufacture. Plasma-driven chemical reactions and/or energetic ion beams are used to remove materials in dry-etching system. The most significant advantage of dry over wet etching is that it provides higher resolution potential by overcoming the problem of isotropy. Other benefits are the reduced chemical hazard and waste treatment problems and the ease of process automation and tool clustering. There are many terms used to describe dry etching, including plasma etching and reactive-ion etching (RIE). Dry etching is the generic term for the process, while, in general, plasma etching can also be used interchangeably. RIE refers to a specific type of dry-etching technique in which ion-assistance is used to obtain anisotropic pattern transfer.

2.2 Dry Etching

2.2.1 Mechanisms of Dry Etching

The predominant advantage of dry over wet etching is anisotropy. It is necessary to understand the range of directionality that may be obtained using dry-etching techniques. There are four basic categories for the mechanisms involved:

2.2.1.1 Sputtering

The interaction is purely physical, involving momentum transfer. The positive ions strike the substrate with high energy. The surface atoms receive the energy transferred from the ions and are ejected from the surface, leading to material removal. Sputtering lacks selectivity among different materials because the ion energy required to eject the surface atoms is very large compared with the differences in surface bond energies and chemical reactivity. Additionally, the etch rates are very slow by sputter etching.

2.2.1.2 Purely chemical

The thermalized neutral radicals react with substrate material and create volatile products which escape from the surface and are pumped out of the chamber. Purely chemical is the most selective mechanism. However, it is usually non-directional (isotropy) and cannot maintain the small dimensions needed in the manufacturing processes. There is little surface reaction with neutral radicals, until energetic ions enhance the reactivity of a substrate or product layer, allowing the formation of volatile products. Its particular advantage is anisotropy.

2.2.1.3 Ion-enhanced energetic mechanism

There is little surface reaction with neutral radicals, until energetic ions enhance the reactivity of a substrate or product layer, allowing the formation of volatile product. Its particular advantage is anisotropy.

2.2.1.4 Ion-enhanced inhibitor mechanism

The inhibitor species form a polymer-like material on the sidewall which helps to block etching of the sidewall and leads to increased anisotropy. The vertical sidewall of features is protected by the materials formed on the sidewall, while ion bombardment suppresses inhibitor growth on the horizontal surfaces that are etched. This phenomenon is called sidewall passivation.

2.3 Dry-Etching Techniques

Dry etching takes place through a combination of chemical and physical components in order to obtain the desired results. The chemical-etch component is rapid but isotropic; the physical component is slow but anisotropic. Therefore, we need to find the best etching technique in which the physical and chemical components are balanced.

Below are some of the dry-etching techniques in common usage.

2.3.1 Ion Milling

This is also called ion etching and is a purely physical process. A typical configuration of an ion-milling machine employs a highly energetic (500–800 eV) inert ion beam (typically Ar^+) to erode the surface by bombardment at a pressure of less than 10^{-4} Torr. The advantage of the technique is the complete absence of undercutting. The disadvantages are that there is little selectivity to mask materials, low etch rates, redeposition on feature sidewalls, and damage to the device due to the high ion energies involved.

2.3.2 Plasma Etching

Plasma is a gas which contains equal numbers of positive and negative charges, as well as neutral atoms, radicals, and molecules. These reactive species serve to chemically etch away materials such as resist, dielectrics, or metals at pressures greater than 100 mTorr. During the purely chemical etching process, three steps occur: adsorption of the necessary species on the materials surface, chemical reaction, and desorption of the products. The advantage of the technique is rapid etch rates, but the drawbacks are isotropy, a tendency for strong loading effects, and the release of heat.

The basic configuration is a parallel plate, planar plasma etching geometry. There is a zone referred to as the plasma sheath that separates the plasma from the sample electrode. The electrodes become negatively charged by the electrons from the plasma because the electrons have greater mobility than do positive ions in the plasma.

2.3.3 Reactive-Ion Etching

The mechanism of operation of this technique is very similar to the plasma-etching technique. To increase etch directionality, lower pressure and a different equipment configuration are employed. It is typically performed at 10–100 mTorr pressure in an asymmetric parallel-plate reactor and is highly anisotropic which means the vertical etch rate far exceeds the lateral etch rate. The area of a flat electrode that holds the wafers is much smaller than the other electrode. The smaller electrode is powered by radio frequency (RF) power (usually 13.56 MHz). The ion density is in the range 10^{10}–10^{11} cm^{-3} for most conventional RF discharges and up to 10^{12} cm^{-3} for some of the new high-density discharge, such as Electron Cyclotron Resonance (ECR). The disadvantage of RIE is convective heating, and this is reduced by putting the wafers on a cooled stage to restrict temperature rise during etching.

2.3.4 Reactive-Ion-Beam Etching

This is a further extension of RIE and could be considered as an intermediate step between RIE and ion milling. Unlike RIE, some physical sputtering can remove the non-volatile products resulting from the chemical reactions. Pressure is usually in the range of 10^{-4} Torr. Compared to ion milling, reactive-ion-beam etching (RIBE) has faster etch rates, better selectivity, and fewer redeposition problems. However, compared with RIE and plasma etching, its throughput is lower.

2.4 Plasma Chemistries

The typical plasma chemistries for dry etching of III-V semiconductors are shown in Table 2.1. For Ga-based compound, Cl_2-based mixtures especially BCl_3 and $SiCl_4$ have been the most popular because of their practical etch rates and smooth morphologies. Generally, pure Cl_2 produces etch rates too high for control in device fabrication, and also it does not attack the native oxide on compound semiconductors, leading to rough surface morphologies where the oxide is not broken-through uniformly. For In-based compounds, the CH_4/H_2 chemistry is attractive since it readily removes In as methyl-related species. The rates are low, but the etching is smooth and anisotropic. Drawbacks include the heavy polymer deposition within the chamber, the need for precise reactor conditioning, and the probability of hydrogen passivation in the near-surface region of the semiconductor. The Br_2 and I_2 chemistries are less commonly used, but I_2 is attractive for its ability to provide high etch rates for In-containing materials. In many cases, an effective compromise is to use Cl_2-based plasmas but with the sample held at elevated temperature in order to increase the volatility of the $InCl_3$ etch product. This can lead to etch rates in excess of 1 micron per minute in high density plasma reactors if the sample is held at 130–150°C.

Table 2.1 Typical etch mixtures for III-V semiconductors

Chemistry	Comments
Cl_2-Based Cl_2, $SiCl_4$, BCl_3, PCl_3, CCl_2F_2	• Usually have addition of Ar, He • Rough etching for In-based compounds • Poor selectivity over photoresist
CH_4-Based CH_4/H_2, CH_4/He, C_2H_6/H_2, C_3H_8/H_2	• Ar often added to enhance stability • Heavy polymer deposition on mask at high pressure or CH_4 content • CH_4 to H_2 ratio must be in range 0.1–0.4 for smooth morphologies
Br_2-Based HBr, CF_3Br, Br_2	• Corrosive to gas lines
I_2-Based HI, CH_3I, I_2	• High rates for In-based compounds • No polymer deposition

During dry etching there are several problems that may occur. First, energetic ion bombardment leads to the creation of point defects that degrade the electrical and optical properties of the material. Second, the composition of the near-surface region may change due to preferential loss of one or more of the elements or there may be residual etch products or polymeric fragments deposited on the surface from the plasma itself or from redeposition of mask material if that degrades during the plasma exposure [1–12].

Due to limited wet chemical etch results for the group-III nitrides, a significant amount of effort has been devoted to the development of dry-etch processing for these materials [1–4]. Dry-etch development was initially focused on mesa structures where high etch rates, anisotropic profiles, smooth sidewalls, and equirate etching of dissimilar materials were required. For example, commercially available LEDs and laser facets for GaN-based laser diodes were patterned using RIE. However, as interest in high power, high temperature electronics increased, etch requirements expanded to include smooth surface morphology, low plasma-induced damage and selective etching of one layer on top of another. Dry-etch development is further complicated by the inert chemical nature and strong bond energies of the group-III nitrides as compared to other compound semiconductors. GaN has bond energy of 8.92 eV/atom, InN 7.72 eV/atom, and AlN 11.52 eV/atom.

2.5 Plasma Reactors

Dry plasma etching has become the dominant patterning technique for the group-III nitrides, due to the shortcomings in wet chemical etching. Plasma etching proceeds by physical sputtering, chemical reaction, or a combination of the two often referred to as ion-assisted plasma etching. Physical sputtering is dominated by the acceleration of energetic ions formed in the plasma to the substrate surface at relatively high energies, typically >200 eV. Due to the transfer of energy and momentum to the substrate, material is ejected from the surface. This sputter mechanism tends to yield anisotropic profiles; however, it can result in significant damage, rough surface morphology, trenching, poor selectivity, and non-stoichiometric surfaces, thus, degrading device performance. The measured sputter rates for GaN, InN, AlN, and InGaN as a function of Ar^+ ion energy increased with ion energy but were quite slow at less than 600 Å/min, due to the high bond energies of the group III-N bond [11].

Chemically dominated etch mechanisms rely on the formation of reactive species in the plasma which adsorb on the surface, form volatile etch products and then desorb from the surface. Since ion energies are relatively low, etch rates in the vertical and lateral directions are often similar, thus, resulting in isotropic etch profiles and the loss of critical dimensions. However, due to the low ion energies used, plasma-induced damage is minimized. Alternatively, ion-assisted plasma etching relies on both chemical reactions and physical sputtering to yield anisotropic profiles at reasonably high etch rates. Provided the chemical and physical component of the etch mechanism are balanced, high resolution features with minimal damage can be realized and optimum device performance can be obtained.

2.5.1 Reactive-Ion Etching

Reactive-Ion Etching RIE utilizes both the chemical and physical components of an etch mechanism to achieve anisotropic profiles, fast etch rates, and dimensional control. RIE plasmas are typically generated by applying RF power of 13.56 MHz between two parallel electrodes in a reactive gas. The substrate is placed on the powered electrode where a potential is induced, and ion energies, defined as they cross the plasma sheath, are typically a few hundred eV. RIE is operated at low pressures, ranging from a few mTorr up to 200 mTorr, which promotes anisotropic etching due to the increased mean free paths and reduced collisional scattering of ions during acceleration in the sheath.

2.5.2 High-Density Plasmas

The use of high-density-plasma etch systems, including ECR, inductively coupled plasma (ICP), and magnetron RIE (MRIE), has resulted in improved etch characteristics for the group-III nitrides as compared to RIE. This observation is attributed to plasma densities which are two to four orders of magnitude higher than RIE, thus improving the III-N bond breaking efficiency and the sputter desorption of etch products formed on the surface. Additionally, since ion energy and ion density can be more effectively decoupled as compared to RIE, plasma-induced damage is more readily controlled. High-density ECR plasmas are formed at low pressures with low plasma potentials and ion energies due to magnetic confinement of electrons in the source region. The sample is located downstream from the source to minimize exposure to the plasma and to reduce the physical component of the etch mechanism. Anisotropic etching can be achieved by superimposing an RF bias (13.56 MHz) on the sample and operating at low pressure (<5 mTorr) to minimize ion scattering and lateral etching. However, as the RF biasing is increased, the potential for damage to the surface also increases.

ICP offers another high-density-plasma etch platform to pattern group-III nitrides. ICP plasmas are formed in a dielectric vessel encircled by an inductive coil into which RF power is applied (Fig. 2.1). The alternating electric field between the coils induces a strong alternating magnetic field trapping electrons in the centre of the chamber and generating a high-density plasma. Since ion energy and plasma density can be effectively decoupled, uniform density and energy distributions are transferred to the sample while keeping ion and electron energy low. Thus, ICP etching can produce low damage while maintaining fast etch rates. Anisotropy is achieved by superimposing of RF bias on the sample. ICP etching is generally believed to have several advantages over ECR including easier scale-up for production, improved plasma uniformity over a wider area, and lower cost-of-operation. Figure 2.2 shows a schematic of the plasma parameters and sample position in a typical high-density plasma reactor.

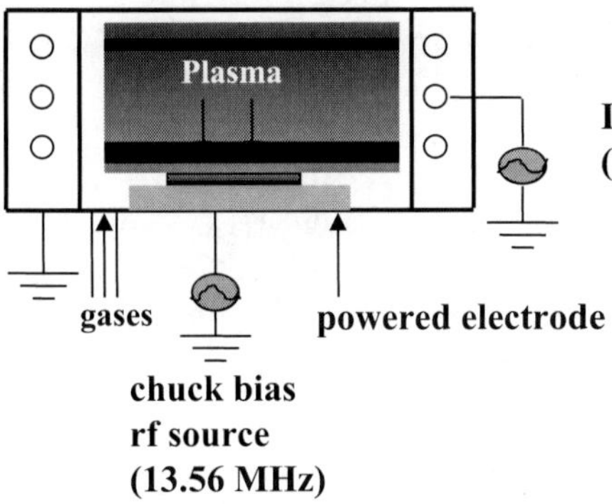

Figure 2.1 Schematic of inductively coupled plasma reactor.

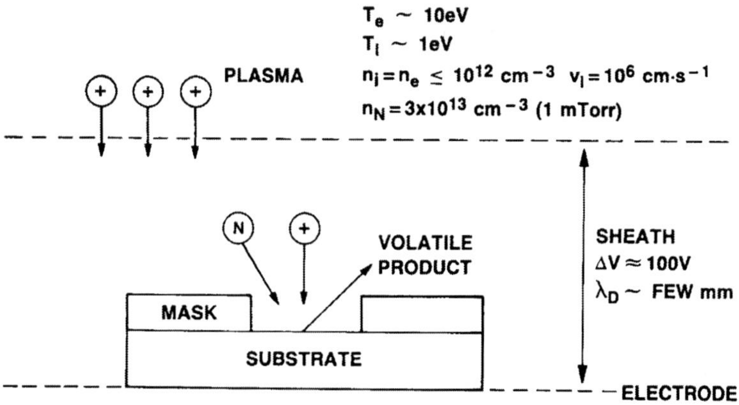

Figure 2.2 Schematic diagram of high density plasma etching process.

MRIE is another high-density etch platform which is comparable to RIE. In MRIE, a magnetic field is used to confine electrons close to the sample and minimize electron loss to the wall. Under these conditions, ionization efficiencies are increased and high-plasma densities and fast etch rates are achieved at much lower dc biases (less damage) as compared to RIE. GaN etch rates of ~3500 Å/min were reported in BCl_3-based plasmas at dc biases < −100 V. The etch was fairly smooth and anisotropic.

2.5.3 Chemically Assisted-Ion-Beam Etching

In chemically assisted-ion-beam etching (CAIBE), reactive gases are added to the plasma downstream of the acceleration grids, thus enhancing the chemical component of the etch mechanism, whereas in RIBE, reactive gases are introduced in the ion source. Both etch platforms rely on relatively energetic ions (200–2000 eV) and low-chamber pressures (<5 mTorr) to achieve anisotropic etch profiles. However, with such high ion energies, the potential for plasma-induced damage exists. GaN etch rates are compared in Fig. 2.3 for RIE, ECR, and ICP $Cl_2/H_2/CH_4/Ar$ plasmas as well as a RIBE Cl_2/Ar plasma. CH_4 and H_2 were removed from the plasma chemistry

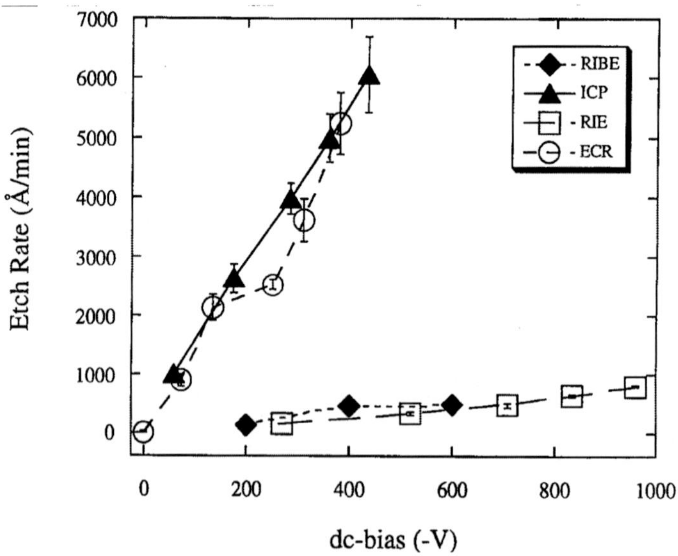

Figure 2.3 GaN etch rates in RIE, ECR, ICP, and RIBE Cl_2-based plasmas as a function of dc bias.

to eliminate polymer deposition in the RIBE chamber. Etch rates increased as a function of dc bias independent of etch technique. GaN etch rates obtained in the ICP and ECR plasmas were much faster than those obtained in RIE and RIBE. This was attributed to higher plasma densities (1–4 orders of magnitude higher) which resulted in more efficient breaking of the III-N bond and sputter desorption of the etch products. Slower rates observed in the RIBE may also be due to lower operational pressures (0.3 mTorr compared to 2 mTorr for the ICP and ECR) and/or lower ion and reactive neutral flux at the GaN surface due to high source-to-sample separation.

2.5.4 Plasma Chemistries

2.5.4.1 Cl_2-based

Etch characteristics are often dependent upon plasma parameters including pressure, ion energy, and plasma density. As a function of pressure, plasma conditions, including the mean free path and the collisional frequency, can change resulting in changes in both ion energy and plasma density. As an example, GaN etch rates are plotted as a function of dc bias (which correlates to ion energy) for an ICP-generated BCl_3/Cl_2 plasma in Fig. 2.4. The GaN etch rates increased monotonically as the dc bias or ion energy increased. Etch rates increased due to improved sputter desorption of etch products from the surface as well as more efficient breaking of the Ga-N bonds. Etch rates have also been observed to decrease under high ion bombardment energies due to sputter desorption of reactive species from the surface before the reactions occur. This is often referred to as an adsorption-limited etch regime. In Fig. 2.5, SEM micrographs are shown for (a) −50, (b) −150, and (c) −300 V dc bias. The etch profile became more anisotropic as the dc bias increased from −50 to −150 V dc bias due to the perpendicular path of the ions relative to the substrate surface which maintained straight wall profiles. However, as the dc bias was increased to −300 V, a tiered etch profile with vertical striations in the sidewall was observed due to the erosion of the resist mask edge. The GaN may become rougher with these conditions due to mask redeposition and preferential loss of nitrogen. GaN etch rates increased as the ICP source power increased due to higher concentrations of reactive species which increases the chemical component of the etch mechanism and/or higher ion flux which increases the bond breaking and sputter desorption efficiency of the etch. Etch rates have also been observed to stabilize or decrease under high plasma flux conditions due to either saturation of reactive species at the surface or sputter desorption of reactive species from the surface before the reactions occur. The etch profile was anisotropic and smooth up to 1000 W ICP power where the feature dimensions were lost and sidewall morphology was rough due to erosion of the mask edge under high-plasma flux conditions.

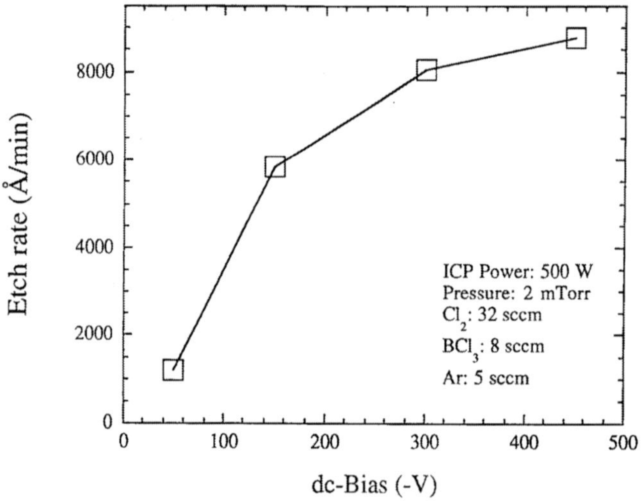

Figure 2.4 GaN etch rates as a function of dc bias in an ICP-generated $BCl_3/Cl_2/Ar$ plasma at 32 sccm Cl_2, 8 sccm BCl_3, 5 sccm Ar, 500 W ICP source power, 2 mTorr pressure, and 10°C electrode temperature.

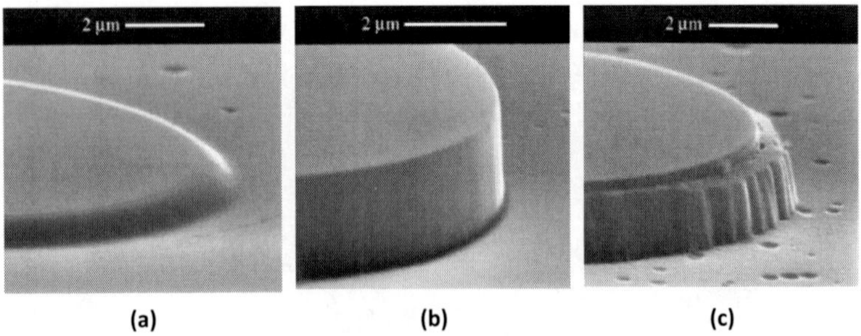

Figure 2.5 SEM micrographs for GaN etched at (a) –50, (b) –150, and (c) –300 V dc bias. ICP etch conditions were 32 sccm Cl_2, 8 sccm BCl_3, 5 sccm Ar, 500 W ICP source power, 2 mTorr pressure, and 10°C electrode temperature.

As established above, etch rates and profiles can be strongly affected by the volatility of the etch products formed. For halogen-based plasmas, etch rates are often limited by the volatility of the group-III halogen etch product. For Ga- and Al-containing films, chlorine-based plasmas typically yield fast rates with anisotropic, smooth etch profiles. CH_4/H_2-based plasma chemistries have also yielded smooth, anisotropic profiles for Ga-containing films, however at much slower rates. Based only on a comparison of etch product volatility, slower etch rates in CH_4-based plasmas is unexpected since the $(CH_3)_3Ga$ etch product has a much lower boiling point than $GaCl_3$. This observation demonstrates the complexity of the etch process where redeposition, polymer formation, and gas-phase kinetics can influence the results. As shown above, etch rates for In-containing films obtained in room temperature chlorine-based plasmas tend to be slow with rough surface morphology and overcut profiles due to the low volatility of the $InCl_3$ and preferential loss of the group-V etch products. However, at elevated temperatures (>130°C), the $InCl_3$ volatility increases and the etch rates and surface morphology improve. Significantly better room temperature etch results are obtained in CH_4/H_2-based plasmas due to the formation of more volatile $(CH_3)_3In$ etch products. It is always necessary to choose plasma conditions that minimize disruption to the etched surface. Figure 2.6 shows an example of an increasingly N-deficient GaN surface as a result of higher ion energy during plasma etching.

The source of reactive plasma species and the addition of secondary gases to the plasma can vary etch rates, anisotropy, selectivity, and morphology. The fragmentation pattern and gas-phase kinetics associated with the source gas can have a significant effect on the concentration of reactive neutrals and ions generated in the plasma, thus affecting the etch characteristics. Secondary gas additions and variations in gas ratios can change the chemical: physical ratio of the etch mechanism. BCl_3/Cl_2 plasmas have shown encouraging results in the etching of GaN films and are often used for via-hole etching in GaAs substrates. The addition of BCl_3 to a Cl_2, plasma can improve sputter desorption due to higher mass ions and reduce surface oxidation by gettering H_2O from the chamber.

The simple Cl_2/Ar chemistry works very well for most device fabrication processes for Ga-based structures, providing controllable etch rates. Even at biases less than 90 V, the GaN etch rate is still typically ~1000 Å·min^{-1} and is 2–4 times higher for GaAs.

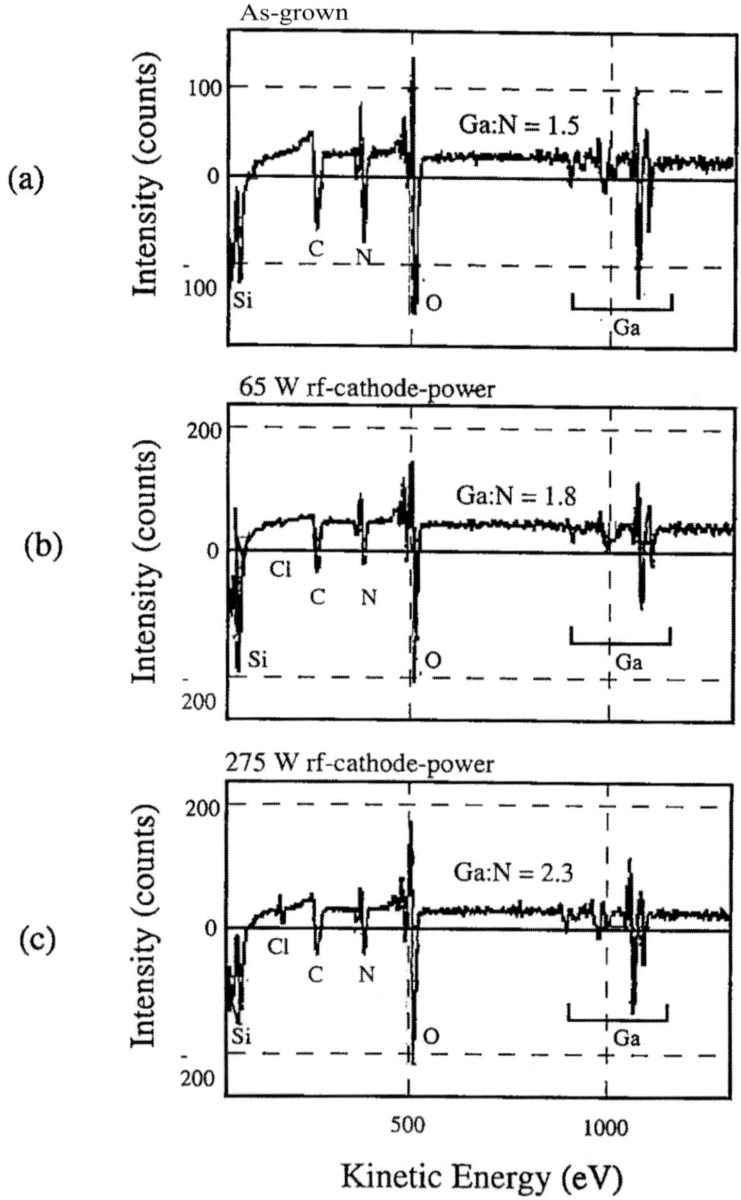

Figure 2.6 AES surface scans of GaN (a) before exposure to the plasma, (b) at 65 W (–120 V bias), and (c) 275 W RF-cathode power (–325 V bias), 1 mTorr, 170°C, and 850 W microwave power in an ECR-generated Cl_2/H_2 discharge.

2.5.4.2 I_2 and Br_2 based

Other halogen-containing plasmas, including HI/Ar, ICl/Ar, IBr/Ar, BBr_3/Ar, and BI_3/Ar, have been used to etch-compound semiconductors with promising results. Etch rates above 1 micron per minute are obtained for InP with iodine-based plasmas in high density systems. GaN etch rates >1.3 μm/min were obtained in the ICl/Ar plasma at a RF power of 250 W (bias of −200 V) while GaN etch rates were typically <4000 Å/min in IBr/Ar. InI_x products have higher volatility than corresponding $InCl_x$ species, making iodine an attractive enchant for InGaN alloys. The inter-halogen compounds are weakly bonded and, therefore, should easily break apart under plasma excitation to form reactive iodine, bromine, and chlorine. Selectivities of >100 can be achieved for both InN/AlN and InN/GaN. Maximum selectivities of ~100:1 for InN/AlN and ~7.5:1 for InN/GaN are obtained. InN etch rates are approximately a factor of 2 faster in BI_3/Ar than in BBr_3/Ar even for lower RF chuck powers. This is expected from taking into consideration the relative stabilities of the respective In etch products (the InI_3 melting point is 210°C; $InBr_3$ sublimes at less than 600°C). The resultant selectivities are ~100:1 for InN over GaN with BI_3, whereas BBr_3 produced somewhat lower values.

2.5.4.3 CH_4/H_2/Ar

This is a universal plasma chemistry for etching III-V semiconductors and is quite robust, being relatively insensitive to background water vapor pressure in the etch reactor. In RIE systems operating at higher pressures (>30 mTorr), there is significant polymer deposition within the chamber and on the mask. It is necessary to remove this polymer by periodically using oxygen plasmas. This must be followed by reconditioning the chamber walls by running the CH_4/H_2 plasma to obtain reproducible etch rates. Typical etch rates for InP in RIE CH_4/H_2 plasmas are a few hundred angstroms per minute, with higher rates for high-density plasmas. It is not possible, at least in our experience, to use very high plasma powers with this mixture because of preferential loss of *P* from the InP surface. ICP GaN, InN, and AlN etch rates approaching 2500 Å/min have been achieved in CH_4/H_2Ar and $CH_4/H_2/N_2$ plasmas. Etch rates increased with increasing dc bias or ion flux and were higher in CH_4/H_2/Ar plasmas. Anisotropy and surface morphology were good over a wide range of conditions. As compared to Cl-based plasmas, etch rates were consistently slower which may make the CH_4/H_2-based processes applicable for devices where etch depths are relatively shallow and etch control is extremely important.

2.5.5 Sidewall Morphology

The morphology of etched sidewalls is especially critical in the formation of laser mesas for ridge waveguide emitters or for buried planar devices. The vertical striations observed in the semiconductor sidewalls may be due to striations in the photoresist mask which were transferred into the semiconductor feature during the etch. Improved sidewall morphology can be achieved using more etch-resistant masks, such as SiO_2 mask. Vertical striations in the SiO_2 mask can be reduced by optimizing the lithography process used to pattern the SiO_2. The SiO_2 is then patterned in a SF_6/Ar plasma in which a low temperature dielectric overcoat is used to protect the resist sidewall during the etch.

2.5.6 Endpoint Detection

Devices such as AlGaAs/GaAs heterojunction bipolar transistors (HBTs) are increasingly needed in the rapidly growing telecommunication industry [5,6]. One issue for plasma etching is to have a reliable process for etching of AlGaAs over GaAs. Etching of GaAs over AlGaAs is also important. However, etching of AlGaAs over GaAs is still an issue [7–9]. There have been no reports on practical plasma chemistry for the selective etching of AlGaAs over GaAs. Precise control of the etch-stop process is extremely important for process reproducibility and device reliability. Automatic control of the process is another issue for mass production. Therefore, it is very useful to have a good technique for *in-situ* end-point detection during plasma etching.

Dry etching with the BCl_3/N_2 chemistry is an attractive choice for etching of AlGaAs/GaAs-based devices [7,8]. It is generally observed that interaction of radicals of BCl_3 and N_2 with photoresist provides excellent sidewall passivation during the etching of GaAs-based materials. In HBT manufacturing, the realization of vertical sidewalls is important for further processing, such as dielectric passivation of the sidewall.

Optical emission spectroscopy (OES) can be an excellent solution as an *in-situ* end-point technique for AlGaAs/GaAs plasma etching, if there is a clear difference of intensity of the monitoring peak when the underlying GaAs layer is exposed [12]. Using an OES technique provides convenience of wafer loading and no sacrifice of real estate on the valuable substrate, which are significant differences compared with using laser reflectometry. Al composition was about 20% in AlGaAs and pattern density of photoresist on the substrates was about 50%. Optical emission spectroscopy was used for detection of plasma emission light. Ga emission peak (wave length is 417 nm) was monitored during etching.

Figure 2.7 shows a Ga peak (light emission wave length is 417 nm) intensity change and a slope of the intensity as a function of time during AlGaAs/GaAs etching. Notice that the intensity was significantly changed when the underlying GaAs was exposed to the plasma. With this OES data, it is not difficult to make a sequence for automatic control of etch depth during AlGaAs/GaAs etching, which will significantly improve reproducibility of the process. The *in-situ* end-point technique provides a more reliable process time and also helps compensate for thickness variations in the AlGaAs epitaxial layer. The intensity of the raw Ga peak continuously increased during AlGaAs etching. The constant increase of the Ga peak intensity is due to the presence of accumulated residual Ga by-products in the plasma and re-etching of the residuals on the chamber wall during AlGaAs etching rather than actual etch rate change of AlGaAs in the wafer.

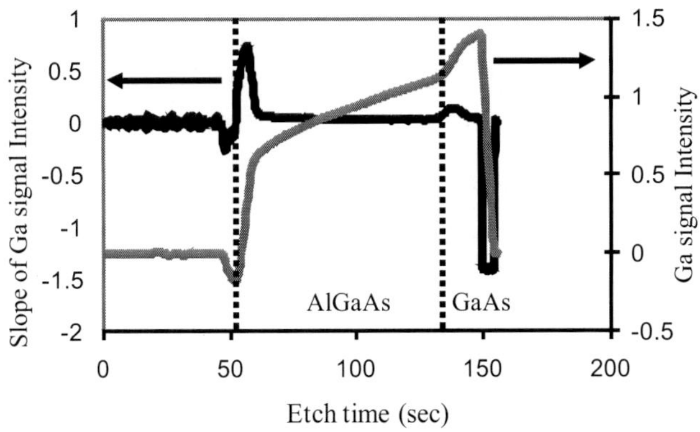

Figure 2.7 Trace of Ga peak intensity (417 nm) and its slope during AlGaAs/GaAs etching in BCl_3/N_2 ICP etching.

Figure 2.8 shows an SEM micrograph after etching of the AlGaAs/GaAs layer in BCl_3/N_2 plasma. Note that the etched surface was extremely clean and smooth and also that the sidewall of the AlGaAs/GaAs layer was quite vertical. Selectivity of etch rate of AlGaAs to the photoresist was about 4:1 at the process condition.

Figure 2.8 An SEM micrograph of AlGaAs/GaAs layer after inductively coupled BCl_3/N_2 plasma etching.

2.5.7 Selectivity

Selectivity of etching rates between two layers occurs because of the differences in volatility of the respective etch products or the formation of an etch-stop material on the surface. An example is the use of Cl_2/SF_6 for selective etching of GaAs over AlGaAs. In this case, selectivity is achieved because of the formation of involatile AlF_3 on the AlGaAs surface after etching through the GaAs. A similar effect can be achieved through the addition of oxygen to chlorine plasma to form AlO_X species on the AlGaAs surface. Since the selectivity occurs through a chemical mechanism, it is a strong function of the incident ion energy and flux and high density plasmas generally show lower selectivites than their RIE counterparts. For example, in general, GaN:AlN and GaN:InN etch selectivities are <10:1 as a function of plasma chemistry for Cl_2- or BCl_3-based plasmas. GaN:AlN and GaN:InN etch selectivities were higher for Cl_2-based ICP plasmas as compared to BCl_3-based ICP plasma due to higher concentration of reactive Cl produced in the Cl_2-based plasmas, thus resulting in faster GaN etch rates. Alternatively, InN and AlN etch rates showed much less dependence on plasma chemistry and were fairly comparable in Cl_2- and BCl_3-based plasmas. Due to the high ion flux in the ICP tools, the usual etch stop reactions are less effective; as an example, the sputter desorption of the AlF_3 may occur prior to passivation of the surface.

2.6 Conclusions

Dry etching of compound semiconductors is widely used for pattern transfer in fabricating optoelectronic devices, such as in forming mesas for wave-guiding in laser diodes. The choice of plasma chemistry is dictated by the materials to be etched. High-density plasma tools offer a great deal of flexibility in designing etch processes because of their wider and nearly independent control of ion flux and ion energy. They can also be operated as RIE tools by simply not powering the high density plasma source and only applying power to the cathode. Wet etching is a simple and effective method for pattern transfer but cannot be used for making very fine features in general because of its anisotropic nature, leading to undercut of the feature sidewalls.

Acknowledgments

The author is very grateful to R. J. Shul, C. Constantine, D. Johnson, J. W. Lee, F. Ren, U. K. Chakrabarti, J. Hong, and C. Vartuli for their collaboration over the years on etching of semiconductors.

References

1. See, for example, the review by Clawson, A.R., Guide to references on III-V semiconductor chemical etching, *Mater. Sci. Eng. R.*, 2001, 31(1–6), 1–438.
2. Pearton, S.J., and Shul, R.J. in *Gallium Nitride I*, edited by J.I. Pankove and T.D. Moustakas (Academic, San Diego, 1998).
3. Shul, R.J. in *Processing of Wide Bandgap Semiconductors*, edited by S.J. Pearton (Noyes, Park Ridge, NJ, 1999).

4. Adesida, A.T. Pin g, C. Youtsey, T. Sow, M. Asif Khan, D.T. Olsen., and J.N. Kuznia, Characteristics of chemically assisted ion beam etching of gallium nitride, *Appl. Phys. Lett.*, 1994, 65, 889–891.

5. See, for example, *Current Trends in Heterojunction Bipolar Transistors*, ed. M.F. Chang. World Scientific, MA (1996).

6. See, for example , Asbeck, P.M., Chapter 6. in *High-Speed Semiconductor Devices*, ed. S.M. Sze, John Wiley & Sons, Inc, NY (1990).

7. Leerungnawarat, P., Cho, H., Hays, D.C., Lee, J.W., Devre, M.W., Reelfs, B.H.,. Johnson, D., Sasserath, J.N., Abernathy, C.R., and Pearton, S.J., Selective Dry Etching of InGaP over GaAs in Inductively Coupled Plasmas, *J. Electron. Mater.*, 2000 29, 586–591,

8. Lee, J.W. Devre, M.W., Reelfs, B.H., Johnson, D., Sasserath, J.N., Clayton, F., Hays, D.C., and Pearton, S.J., Advanced selective Dry etching of GaAs/AlGaAs in high density inductively coupled plasmas, *J. Vac. Sci. Technol.*, 2000, A 18, 1220–1226.

9. Salimian, S., and Cooper, C.B. III, Equal rate and anisotropic reactive ion etching of GaAs/AlGaAs heterostructures in $SiCl_4$ plasma, *J. Electrochem. Soc.*, 1989, 136, 2420–2425.

10. Lee, J.W., Hong, J., Lambers, E.S., Abernathy, C. R , Pearton, S.J., Hobson, W.S., and Ren, F., Plasma etching of III-V semiconductors in BCl_3 chemistries: Part I: GaAs and related compounds, *Plasma Chem. Plasma Proc.*, 1997, 17, 155–163.

11. Maeda, T., Lee, J.W., Shul, R.J., Han, J. , Hong, J. , Lambers, E.S. , Pearton, S.J., Abernathy, C.R., and Hobson, W.S, Inductively coupled plasma etching of III-V semiconductors in BCl_3-based chemistries, I. GaAs, GaN, GaP, GaSb and AlGaAs, *Appl. Surf. Sci.*, 1999, 143, 174–179.

12. Collot, P. , Diallo, T, and Canteloup, J., Dry-etch monitoring of III-V heterostructures using laser reflectometry and optical emission spectroscopy, *J. Vac. Sci. Technol. B*, 1991, 9, 2497–2501.

Chapter 3

Nanoscale Split Ring Resonator-Based Metamaterials: Fabrication Techniques, Properties, and Applications

Basudev Lahiri,[a,c,*] Scott G. McMeekin,[b] Richard M. De La Rue,[a] and Nigel P. Johnson[a]
[a] *Optoelectronics Research Group, School of Engineering, University of Glasgow, Glasgow G12 8LT, UK*
[b] *School of Engineering and Computing, Glasgow Caledonian University, Glasgow G4 0BA, UK*
[c] *Present address: Center for Nanoscale Science and Technology, National Institute for Standards and Technology, Gaithersburg, MD-20899, USA*
*basudev.lahiri@nist.gov

Metamaterials are artificial materials that are known to produce extraordinary electromagnetic responses due to their constituent artificially-engineered micro- or nanostructures with dimensions smaller than that of the wavelength of light. The Split Ring Resonator (SRR) is one such nanostructure that forms the basic unit of a metamaterial. Since the dimensions of the SRRs are required to be smaller than the resonance wavelength, their fabrication becomes critical when a response is required at near infrared and optical wavelengths. In this chapter, the various properties of nanoscale

Vistas in Nanofabrication
Edited by Faiz Rahman

ISBN 978-981-4364-56-0 (Hardcover), 978-981-4364-57-7 (eBook)
www.panstanford.com

SRRs are studied that resonate in the infrared and visible spectrum. The SRRs discussed in this chapter are made of aluminium (Al) and gold (Au) and have been fabricated on silicon and silica substrates using electron beam lithography (EBL) nanofabrication techniques. It is shown that, by using aluminium-based SRRs instead of gold ones, the magnetic resonance of SRRs can be shifted well into the visible spectrum. Finally, it is shown that, by adding asymmetry to similar sized geometrical structures such as doubly-split SRRs, it is possible to produce a steeper resonance response, thereby increasing the quality factor of the SRRs. This steep response of asymmetric split ring resonators (A-SRRs) is utilized for the optical detection of very thin films of organic compounds.

3.1 Introduction

Metamaterials are artificially created or engineered materials that typically have periodically arranged features of dimensions smaller than that of the wavelength of light. They demonstrate electromagnetic properties quite different from those of naturally occurring materials. Metamaterials derive their properties from a combination of composition and structure, rather than from their composition alone.

Metamaterials exhibit electromagnetic responses not readily available in naturally occurring materials and structures, such as negative refractive index [1], artificial magnetism [2], super focusing [3], and electromagnetic invisibility (cloaking) [4]. They typically contain fabricated micro or nanostructures (often made of metals) smaller than the wavelength of radiation in the region of interest, which is the guiding factor behind their unusual response to electromagnetic waves. The online encyclopaedia, Wikipedia, provides a huge list of potential applications of metamaterials, including species detection, infrastructure monitoring, battlefield communication, and public safety.

The engineered structures of metamaterials replace the attributes of atoms and molecules that determine the overall electromagnetic response of a conventional material [2,4–5]. Because of the smallness of the structuring in comparison with the wavelength of the light, the incident electromagnetic radiation encounters a nearly homogeneous material during wave-material interaction. The electrons oscillating in highly conductive metals, such as gold (Au), aluminium (Al), silver (Ag), or copper (Cu), give rise to the specific electromagnetic response of the metamaterial, in conjunction with the geometries of the metallic features. By adjusting the geometries of the metallic micro/nanostructural units, customized electromagnetic responses can be obtained [4–5]. Split-ring resonators (SRR) are arguably one of the most common metamaterial micro/nanostructural units and are utilized for their resonant response and negative magnetic properties [4]. The SRRs can be fabricated over a wide range of dimensions to suit applications over a large portion of the electromagnetic spectrum, ranging from radio and microwave frequencies (wavelength 100 km to 1 cm) to the infrared and optical frequency range (wavelengths in the range 10 μm to 400 nm).

3.2 The Split Ring Resonator

In 1999, Pendry *et al.* [2] proposed that by creating artificial structures of dimensions smaller than the wavelength of light in a periodic manner one can obtain negative refractive-index behaviour. These materials were subsequently named "metamaterials" or "Left-handed materials (LHM)". Pendry proposed an array of homogeneously-spaced sub-wavelength SRRs as metal hoops, which as a whole behave like a composite material. These SRRs can act as an LC oscillating circuit containing a magnetic coil of inductance L and a capacitor of capacitance C. In their fundamental resonance, they behave as an LC oscillatory circuit that contains a single-turn magnetic coil of inductance in series with a capacitance produced by the gap between the arms of the SRR. The SRR and its equivalent circuit are illustrated in Fig. 3.1.

The unit cell of a SRR and its two different polarization-dependent electromagnetic responses are shown in Fig. 3.2. For normal incidence with TE polarization mode, when the incident electric field is across the gap (parallel to the X-axis), the electric field couples with the capacitance of the SRR and generates a circulating current around it. This circulating current induces a magnetic field in the base of the SRR and also interacts with the external field to generate the magnetic resonance that can be appropriately called an LC resonance as identified by Linden *et al.* [3]. When the incident light is with TM polarization mode, with the electric field perpendicular to the gap (parallel to Y-axis), the electric field cannot couple to the capacitance of the SRR and, therefore, generates only electric (Plasmon) resonances [2–4].

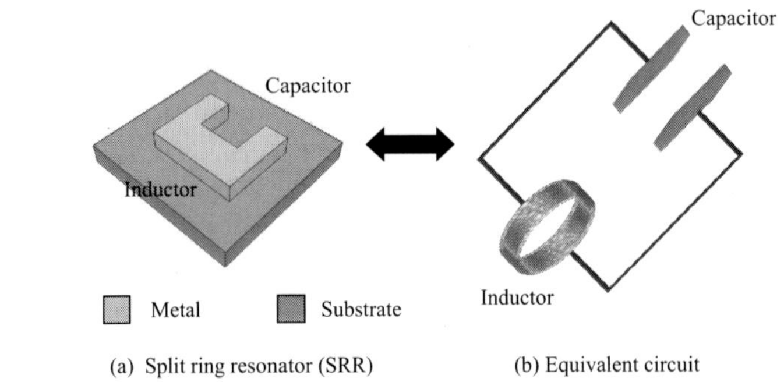

Figure 3.1 A single unit cell of the SRR array and its equivalent LC circuit. The gap of the SRR acts like a parallel plate capacitor, whilst its base acts as an inductor. See also Color Insert.

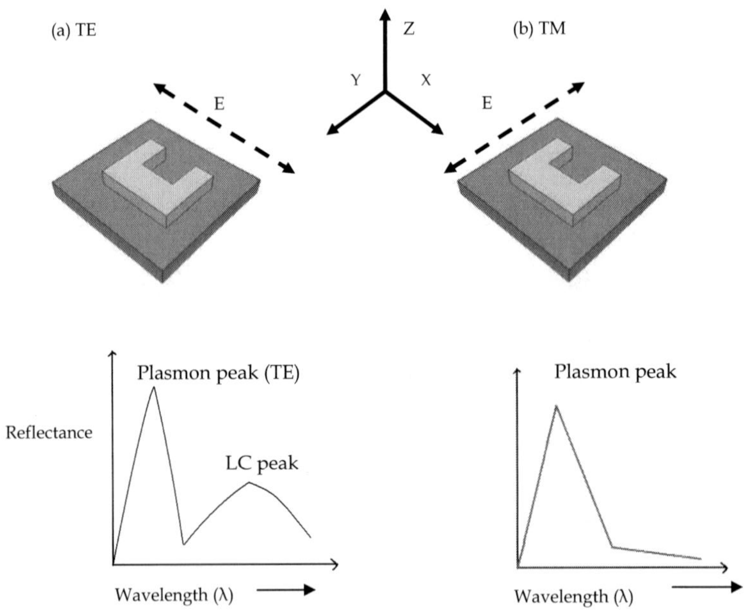

Figure 3.2 Schematic of single SRRs and their corresponding reflectance spectra (Illustration only) with (a) TE measurements of normal to surface incidence having the electric field (dashed line) parallel to the SRR arms (parallel to the X-axis) and (b) TM measurements of normal to surface incidence having the electric field perpendicular to the SRR arms (parallel to Y-axis). See also Color Insert.

It should be noted here that there exists a difference in the spectral position as well as the width of the plasmon peaks obtained from the two different polarization configurations. This has been explained by Enkrich *et al.* [6]. For TE polarization, only the base of the SRR is excited and produces the TE plasmon peak. For TM polarization, both arms of the SRR, which are coupled via its base, are excited. As a result, there is the formation of two new symmetric and asymmetric oscillation modes. The asymmetric mode cannot be excited for symmetric SRRs (see later sections for more details on the asymmetric modes), but the symmetric mode gets excited and is slightly red-shifted as compared to the uncoupled plasmon resonance of the base in TE measurements. The plasmon resonance of the TM mode, hence, has a larger effective electric dipole moment [6].

At the same time, negative permittivity can be achieved by placing a periodic array of thin metallic wires in close proximity. The wires individually act as electric dipoles, and their overall response can give rise to negative permittivity. Combining wires with SRRs can give rise to negative refractive index. In 2001 Smith *et al.* [7] created interlocking units of fibre-glass sheets imprinted with copper rings and wires. This structure experimentally confirmed negative refractive index in microwave region and bent the incident wave the "wrong way" when emerging from the metamaterial [7].

3.2.1 The Magnetic Response of SRRs

As described previously, the SRR acts as an LC circuit, and its magnetic response (LC resonance) is due to the circulation of electric current when the electric field of the normally incident wave is parallel to its gap. Hence, the LC resonance is polarization-dependent. A single SRR is shown in Fig. 3.3 once again with its dimensions that are used to derive the formula for its magnetic resonance (LC resonance).

From basic textbook formula of capacitance of a parallel plate capacitor with standard approximations as given in reference [8], the capacitance (C) of an SRR is calculated as

$$C \propto \frac{\text{Area of the plate}}{\text{Distance between the plates}} \tag{3.1}$$

$$\text{or } C = \varepsilon_0 \varepsilon_c \frac{wh}{\delta} \tag{3.2}$$

where, ε_c is the dielectric constant of the substrate.

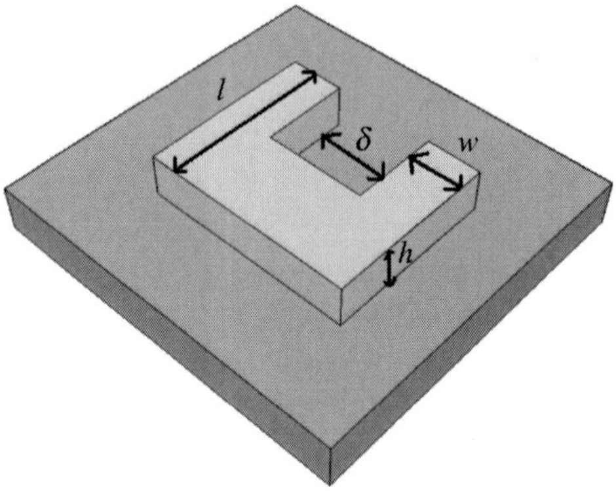

Figure 3.3 A single unit cell of the SRR array, with l being the overall length, w the width of the arm, h the thickness (height) of SRR, and δ the gap between the arms. See also Color Insert.

Similarly, the inductance (L) of a coil of N turns, where $N = 1$ is given as [10].

$$L \propto \frac{\text{Area of coil}}{\text{Length}} \tag{3.3}$$

$$\text{or } L = \mu_0 \frac{l^2}{h} \tag{3.4}$$

Now, the resonance frequency (ω_{LC}) of an LC circuit is given by [10].

$$\omega_{LC} = \frac{1}{\sqrt{LC}} \tag{3.5}$$

Putting the values of C and L from Eqs. 3.2 and 3.4 into Eq. 3.5 we get the LC resonance frequency (ω_{LC}) as [10].

$$\omega_{LC} = \frac{1}{l}\sqrt{\frac{\delta}{w}}\frac{c_0}{\sqrt{\varepsilon_c}} \tag{3.6}$$

The LC resonance wavelength (λ_{LC}) is given as [9].

$$\lambda_{LC} = \frac{2\pi c_0}{\omega_{LC}} = 2\pi l\sqrt{\varepsilon_c}\sqrt{\frac{w}{\delta}}\ \alpha \text{ size} \tag{3.7}$$

Thus, from Eq. 3.7, it could be said that the LC resonance wavelength (λ_{LC}) is proportional to the size of the SRR.

From Eq. 3.7, it is seen that the gap between arms (δ) is very important since if the gap of the SRR is closed, i.e., $\delta = 0$, the LC resonance frequency (ω_{LC}) becomes zero and the LC resonance wavelength (λ_{LC}) becomes infinite [8]. Also from Eq. 3.7, the thickness/height (h) of the SRR plays no significant role in the determination of its LC resonance frequency.

Thus, it is seen that the response of an SRR is dependent only on its size, i.e., dimensions and not on metal properties (and is constant for a given substrate (ε_c) for the surrounding substrate permittivity). As a result, there was a sustained effort in the scientific community to scale the sizes of SRRs to smaller (nanoscale) dimensions in order to push the LC resonance to higher frequencies. It was considered novel work as there was no such magnetic material exhibiting magnetic resonances at such high frequencies. At low frequencies, up to several terahertz, the magnetic resonance (LC resonance) scales reciprocally with dimensions of SRRs. However, at higher frequencies, this scaling law is no longer valid and the response of an SRR no longer depends on its size. It was explained that at higher frequencies (optical and near infrared regimes), the kinetic energy of the electrons in the metal becomes predominant and this in turn kills the circulating electric current produced by the coupling of the SRR arms, in turn destroying its LC response [9–10]. This phenomenon along with some other novel responses of the SRR is discussed in detail in subsequent sections.

3.3 Fabrication of SRR Structures

For the fabrication of SRR structures over dielectric substrates, the process of EBL is used. EBL is ideal for fabricating the extremely small features of SRRs. Most of the EBL tools commercially available offer a resolution of about 30 nm which is suitable for making such structures. In EBL, the exposing energy comes from accelerated electrons. With the help of an electron gun, a beam of electrons with appropriate current density is generated. The electron gun utilizes a thermionic emission coil for the production of electrons. Newer machines make use of field-emission guns to generate the electron beam. Condenser lenses at specific position are used to focus the electron beam to a spot size of 10–25 nm [11]. With computer control, the electron beam could be made to strike any location on the sample stage. In EBL techniques, the pattern to be written over a dielectric substrate (e.g., silicon, silica etc.) is first designed within a CAD software environment, e.g., Tanner L-Edit™. The design is then converted to a GDS file which is readable by the pattern generator connected to the EBL machine. The electron beam is then focused onto polymers called electron-beam resists that are coated over the substrate. Electron-beam resists are electron-sensitive polymers. They are used to cover the entire substrate which is subsequently exposed to the high-energy electron beam so that the desired pattern can be generated into it. The development of an electron resist in a developer solution then produces a resist pattern on the wafer that defines which areas of the wafer will be exposed for material deposition or removal. Electron resists are classified into positive and negative electron resists depending on how they respond to irradiation. For a positive electron resist, irradiation with electrons causes the resist polymer in the exposed areas to be broken into shorter molecular fragments. This results in the formation of low-molecular-weight polymer chains that are later dissolved and removed by the developer solution. For a negative electron-beam resist, the irradiation causes polymer cross-linkage in the exposed areas, resulting in the creation of high-molecular-weight polymer chains that resist dissolution in most solvents. In all of the work described in this chapter, the electron-beam resist used was poly-methyl methacrylate (PMMA) which is a positive-tone electron-beam resist.

In this section, a general description for the fabrication of nanoscale SRRs is provided in order to give the reader a general idea of the steps required. The sample is then prepared as follows.

(i) A Silicon wafer of dimensions around 2 cm × 1 cm is rinsed with acetone in an ultrasonic bath for ~55 seconds then Methanol in the same way for ~55 seconds and then in the same way with isopropyl alcohol (99%) (IPA), for ~55 seconds.

(ii) An e-beam positive resist, typically 4% 2010 PMMA, is then spun over the sample at 5000 rpm for 60 seconds to obtain a resist thickness of 110 nm.

(iii) The sample is then put into a 180°C oven for 15 minutes.

(iv) The sample is then spun with 4% 2041 PMMA at 5000 rpm for 60 seconds to give an obtain an additional resist lower with a thickness of 126 nm. The sample is then baked in the 180°C oven for 1 hour so that the resist can better adhere to the wafer.

(v) The sample is then submitted for E-beam lithography. The dose is varied from ~800–5200 μC/cm^2 with a 100 KeV beam, depending on the feature size required.

(vi) The resolution is selected within the range of 1.25 nm to 0.50 nm.

(vii) The written sample is then developed with 2:1 MIPBK at 23°C for 45 seconds, and cleaned with IPA.

(viii) The sample is then coated with a thin metal film. Typical examples of metal deposition are 2 nm Ti + 48 nm Au or 50 nm Al.

(ix) Excess metal and resist is then lifted off using warm acetone.

To summarize, the pattern is designed and then patterned by EBL. The Structure is formed during metal lift-off. The patterns are written over an area of ~300 μm × 300 μm. All the steps (i) to (ix) are illustrated in the illustration of Fig. 3.4.

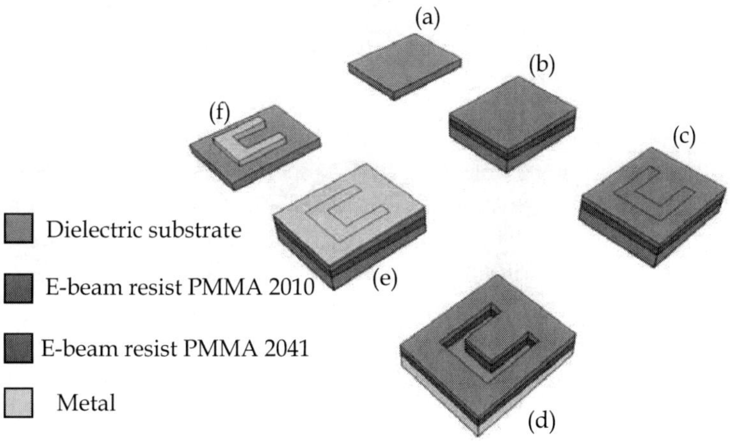

Figure 3.4 Fabrication processes of SRR structures (a) cleaning of silicon substrate; (b) spinning of E-beam resist; (c) E-beam exposure; (d) development; (e) Metal deposition; and (f) Lift-off in warm acetone. See also Color Insert.

A scanning electron microscope (SEM) image of SRRs fabricated using the above process is shown below in Fig. 3.5.

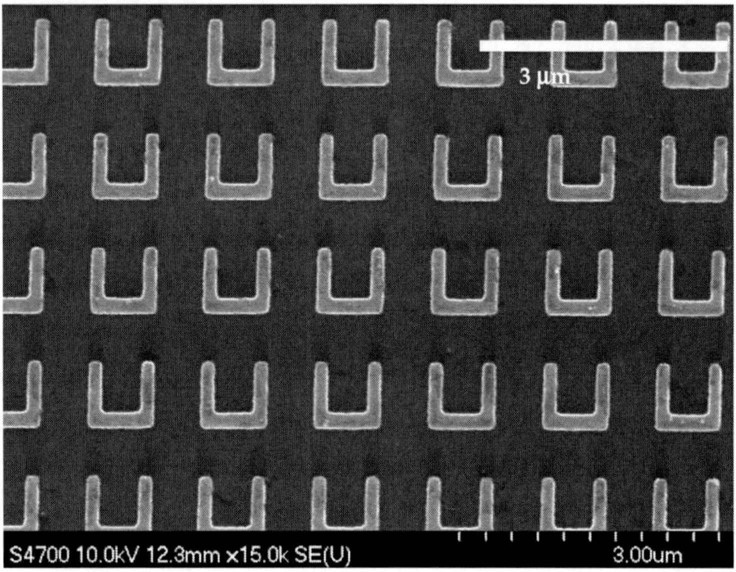

Figure 3.5 Scanning Electron Microscope (SEM) image of gold SRRs fabricated on a silicon substrate.

3.3.1 Modelling and Reflectance Measurements

Computer modelling of the structures was performed using Fullwave, a commercial FDTD simulation package from RSoft. A unit cell was defined with dimensions equal to the spacing of the SRR array, with the SRR structure centred within the cell. The boundary conditions for the cell were set as periodic in the coordinates perpendicular to the propagation direction. This effectively extended the array to infinity, and, as a perfectly matched layer (PML) in the direction of propagation, to eliminate any reflections from the boundaries. The frequency-dependent complex dielectric properties of the metals were modelled using a Lorentz–Drude model, the SiO_2 substrate was modelled using a Lorentz model, and the Si substrate model was based on experimental material parameters measured across the wavelength range of interest [12].

Typically, for the work described in this chapter, the reflectance measurements were performed at normal incidence with an optical microscope, using a ×10 lens with a numerical aperture (NA) of 0.25. A white light source and a monochromator with either an InGaAs detector (for operating range 0.8–1.6 μm) or a SiGe detector (for operating range 0.4–0.8 μm) and a lock-in amplifier were also used. The polarization control in this case was obtained by rotating the samples by 90° with respect to a linearly polarized beam, to obtain two different sets of polarization-dependent data. For reflectance measurements at longer wavelengths (in the range from 2.5 μm to 15 μm), we used a Nicolet Continuum 6700 Fourier transform infrared (FTIR) spectrometer equipped with an IR microscope. A nitrogen-cooled mercury–cadmium–telluride (CMT) detector was used in these measurements. All measurements were performed at normal incidence. This FTIR beam was polarized using a Continuum ZnSe IR polarizer. The reflectance measurements were normalized with respect to the reflectance of a gold mirror and were taken for two orthogonal linear polarizations of incident light oriented simultaneously with respect to the array axes and the SRR element orientation.

Figure 3.6 shows the SEM image of an array of aluminium-based SRRs fabricated on a silicon substrate and its corresponding reflectance spectra measured by the above-mentioned method.

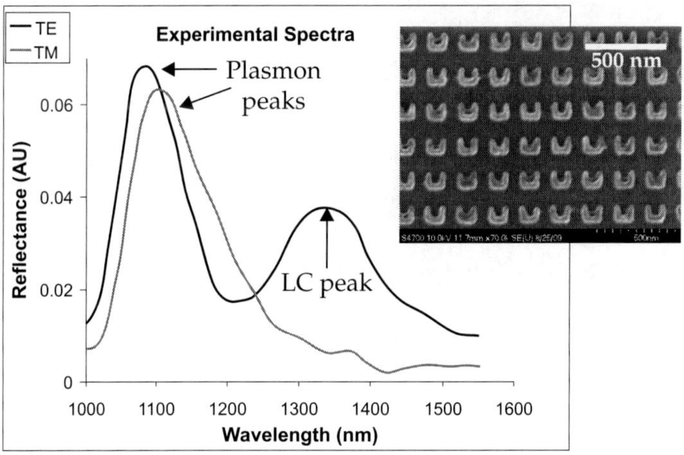

Figure 3.6 SEM image (Inset) of aluminium-based SRRs fabricated on medium-doped silicon substrate and its corresponding normal-to-plane experimental reflectance spectra. The black curve is for TE polarization and the red curve is for TM polarization. The LC peak for TE polarization is observed at 1344 nm and the plasmon peaks for TE and TM polarizations are observed around 1100 nm.

3.4 High Frequency Saturation of SRR

It has previously been found that the magnetic response of the "rectangular C" shaped SRR structures typically saturates just before the visible region is reached and cannot be shifted indefinitely to higher frequencies by reducing the size of the SRRs [13–14]. Tretyakov has studied this phenomenon in the scaling of SRRs and the saturation of the frequency response, using an LC equivalent circuit model [15]. This model considers the dependence of the achievable resonant frequency on the geometric shape of the SRR, as well as arguing for the involvement of an additional kinetic inductance term and an additional distributed capacitance term. In the present section, we discuss the differences that occur in the responses of SRRs with nominally identical dimensions but made up alternatively of aluminium and gold. We report the observation of a substantial shift in response between the aluminium and gold SRRs. Furthermore, we utilize this shift to bypass the resonance saturation behaviour of gold (and, by implication, silver) SRRs and demonstrate a polarization- dependent LC resonance peak at visible wavelengths for aluminium-based SRRs. The changes in the properties of metals at higher frequencies, where the metal properties behave according to the Drude model and both the real and the imaginary parts of the dielectric function must be considered, have so far not been fully explored in relation to the high-frequency resonance saturation behaviour of SRRs [4,13–14].

3.5 Effects of Different Metals on SRR Response

Apart from their physical size, the frequency of the LC resonance of an array of SRRs is also determined by the refractive index of the substrate supporting the SRR patterns [5, 16]. The SRRs fabricated on lower refractive-index materials (e.g., silica) characteristically show their LC resonant response at higher frequencies than those fabricated on silicon.

Figure 3.7 shows various, nominally similar, SRRs made of gold and aluminium, fabricated on silica substrates and their corresponding reflection spectra, displaying LC peaks in the visible part of the spectrum.

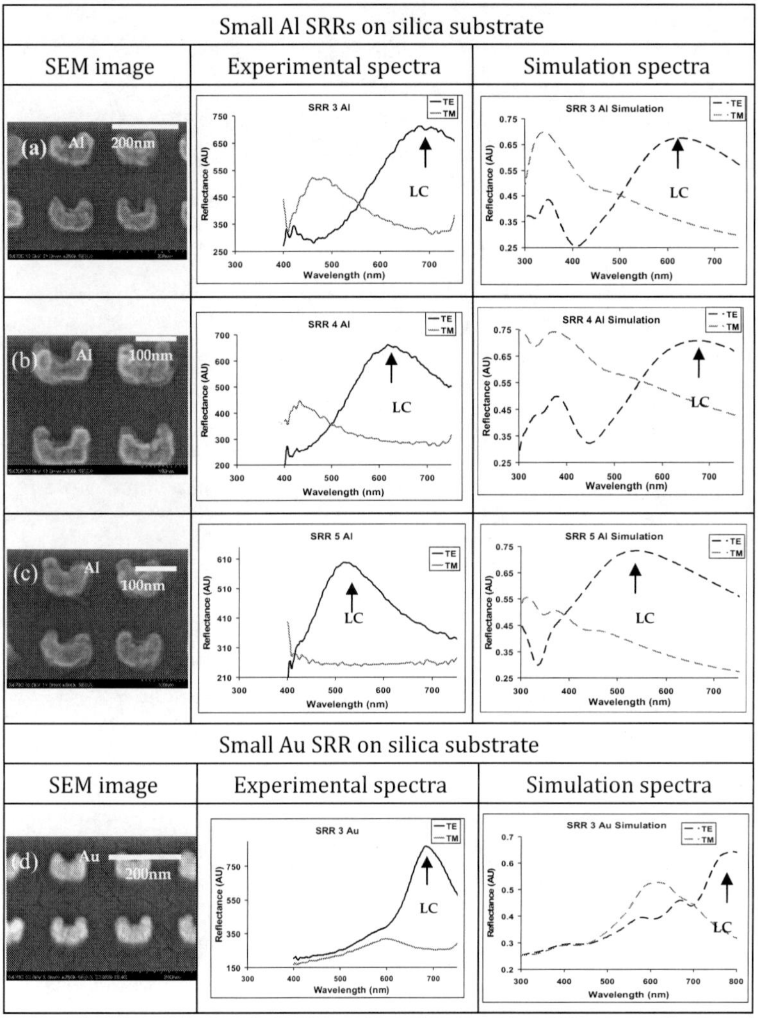

Figure 3.7 The reflectance spectra of the different SRR patterns fabricated on silica. The first three rows are small-sized (~150 to 120 nm) aluminium SRRs with (a) SRR 3 Al (b) SRR 4 Al, and (c) SRR 5 Al, and the last row is small-sized (~120 nm) gold SRR with (d) SRR 3 Au along with their corresponding experimental (solid curves) and simulation (dashed curves) spectra. Black curves (solid and dashed) are TE and red curves (solid and dashed) are TM measurements.

As shown in Fig. 3.7, the aluminium-based SRRs of dimension (*l*) varying from ~150 nm to ~100 nm, fabricated on silica, display the LC resonance from 680 nm to as low as 530 nm wavelengths, well inside the visible region. Whereas the smallest dimension (l) of ~100 nm gold-based SRRs fabricated on silica display the LC peak at 720 nm, just on the edge of the visible spectrum. This clearly demonstrates the advantage of aluminium-based SRRs in pushing the magnetic resonance (LC resonance) to higher frequency in the visible spectrum, when compared with their gold-based counterparts.

Overall, apart from the broadening of the LC feature and the possible reduction in the LC resonance magnitude associated with aluminium, there occurs a substantial shift in the position of the LC and Plasmon peaks between Al and Au SRRs of similar dimensions. According to the Tretyakov model, at higher frequencies an additional inductance term has to be added while interpreting the resonance responses of metallic SRRs. This is due to the additional kinetic inertia associated with the motion of the electrons, which depends on the choice of the metal.

According to the Tretyakov model [15], if the length of the SRR is l, the width of its arm is w and the thickness (height) of the SRR is h and δ is the gap between the SRR arms (as shown in Fig. 3.3) and further assuming uniform current density over the cross section of the SRR, the LC resonant frequency of the SRR is given by:

$$\omega_{LC} = \frac{1}{\sqrt{(L+L_{add})(C+C_{add})}} \tag{3.8}$$

where the additional capacitance and inductance are given by:

$$C_{add} = \frac{\varepsilon_0 \varepsilon_r wh}{l_{eff}} \text{ and } L_{add} = \frac{l_{eff}}{\varepsilon_0 wh\omega_p^2} \tag{3.9}$$

where l_{eff} is the effective conductor length that Tretyakov [10] assumes to be $l_{eff} = (\pi/2)l$ and ω_p is the plasma frequency. The additional inductance term, L_{add}, associated with the kinetic energy of the electrons depends strongly on the plasma resonance frequency and, therefore, on the choice of the metal. This has been shown in previous work [17] where similar-sized aluminium-based SRRs showed LC resonance at higher frequencies than their equivalent gold counterparts. At frequencies below the plasma frequency, it is the strongly plasma-frequency-dependent term for the kinetic inductance that plays a crucial role. Equation 3.9 indicates that this additional inductance is inversely proportional to the square of the plasma frequency of the metal. The plasma frequency of the metal is equal to 3750 THz for aluminium and 2175 THz for gold [18].

The higher plasma frequency of aluminium will result in a lower additional inductance (L_{add}) and, therefore, in a shorter resonance wavelength.

The reduction in the magnitude of the reflection resonance for aluminium SRRs can be attributed to the difference in the damping frequency of the metals. In the Drude model, the damping (collision) frequency is associated with absorption and losses. The damping frequency of aluminium is 19.4 THz and that of gold is 6.5 THz [18]. A metal having higher values of damping frequency tends to be more absorptive and, thus, lossy. Since aluminium has a higher damping frequency than gold, it is more absorptive and hence the magnitude of the response of an array of aluminium SRRs is lower than that of gold. In Fig. 3.8, we display TE field plots of SRR-4Al at both the plasmonic and LC resonant peaks.

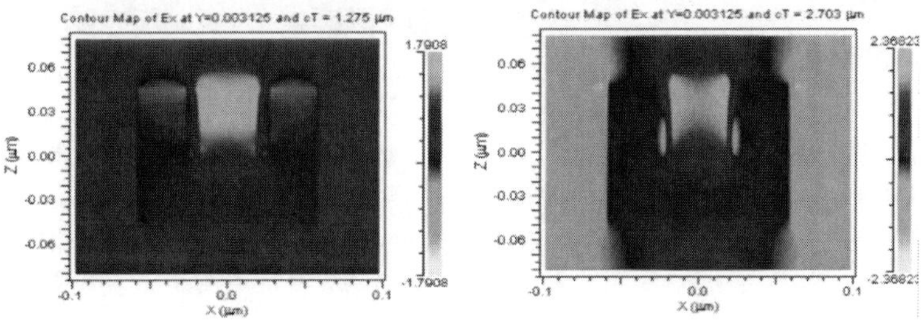

(a) (b)

Figure 3.8 (a) Field profile of SRR-4 Al at the plasmonic resonance peak (370 nm). (b) Field profile of SRR-4 Al at the LC resonance peak (670 nm). See also Color Insert.

In Fig. 3.8, the different distributions of the optical electric field for the two different resonances are shown. In the case of the plasmon resonance, the phase and amplitude of the optical electric field has a uniform distribution around the SRR structure and it is in phase with the field in the gap of the SRRs. The LC resonance on the other hand exhibits a completely different situation, in which the optical electric field between the arms of the SRRs is in anti-phase with the light on either side of the SRR structure. From this difference in behaviour, it may be inferred that the LC resonance condition is a fundamentally different one from the plasmonic resonance condition. This description of the resonance behaviour is consistent with that of Linden *et al.* [4].

3.6 Asymmetric Split Ring Resonators

In the previous section, it was seen that metamaterials based on nanoscale metallic SRRs can be used to produce resonant responses in the visible and infrared wavelength regions. However, the quality factor of resonance associated with metamaterials is usually low due to the strong coupling of the SRRs with free space and that result in radiation losses [19]. This lossy nature of SRRs and its broad resonant response makes it difficult for practical and efficient implementation of metamaterials in applications such as optical sensors. A high quality factor is desirable if the structures are used as sensors, because a shift in a sharp peak is easier to detect than a shift in a gentle slope. It has been shown in a few theoretical studies that it is possible to obtain higher quality factor resonance in metamaterials by introducing certain asymmetries into the SRR geometry [20–22]. Zheludev *et al.* [19] have experimentally demonstrated higher quality factor for metamaterials at microwave frequencies by using A-SRRs.

The asymmetries between the two arms of the double SRR structures produce distinct plasmon resonances that are modified by the presence of the other arm, leading to a steep gradient of the reflection versus frequency curve between the resonances [19–21]. A shift in a sharp feature is more easily identified than a rounded feature and, therefore, increases the sensitivity compared with similar but symmetric SRRs (S-SRRs). The A-SRRs described in this section are composed of two separate metallic arcs of unequal lengths that share the same centre of curvature. The two arcs interact to produce a very steep slope in the reflection spectrum. The A-SRRs described here possess a higher quality factor for their plasmon resonances (a factor of two or more) than the plasmon resonance of its symmetric double or single split counterparts.

The A-SRRs described here have a diameter of 1.2 to 1.5 μm with a strip width of 100 nm and are fabricated in thin gold films on fused silica substrates. The characteristic resonances of such structures are in the mid-infrared (typically at wavelengths in the 3–6 μm) range. The main objective in fabricating such A-SRRs was their potential for high-sensitivity optical detection of organic and bio-chemical substances. Because of this, gold as a metal was chosen to form the A-SRRs due to its good resistance to oxidative corrosion when coming in contact with organic compounds. For the same reasons, fused silica was chosen for its inertness to most organic compounds.

The unit cell of a single A-SRR, as shown in Fig. 3.9, consists of two distinct and opposite single arcs separated by two identical narrow gaps. These two arcs (left hand and right hand) act as two separate resonators, with their individual resonance response depending on their respective length as well as on the angular gap between them. In our analysis, we have varied the angular gap (θ) for the A-SRRs from ~15° to 55°, in steps of 10°, by keeping the length of the left-hand arc constant and shortening the length of the right-hand arc. All the other parameters are identical.

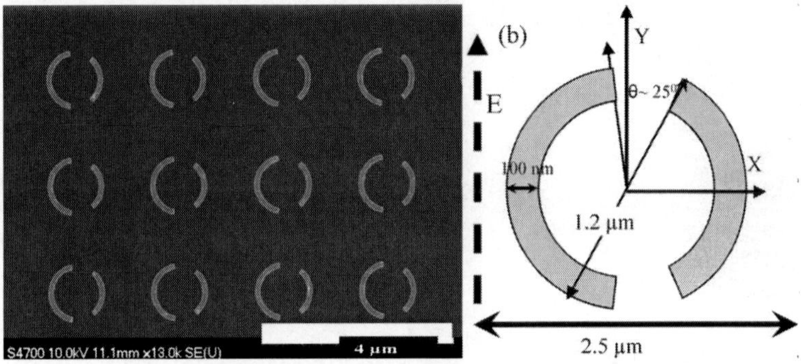

Figure 3.9 (a) SEM micrograph of A-SRR with diameter 1.2 μm and 100 nm strip width. (b) A single unit cell of A-SRR with optimized dimensions. All measurements were taken at normal incidence, with the orientation of the E-field parallel to the Y-axis.

Figure 3.10 depicts the resonant responses of arrays of symmetric and asymmetric SRR arrays, obtained by measuring their corresponding reflectance spectra.

In each case the measurements were made with light at normal incidence (i.e., the Z direction) with the E-field maintained parallel to the Y direction. In the case of S-SRR1, the sizes of the left-hand and right-hand arcs are nominally identical (symmetric), their respective responses are superimposed, giving the overall resonance observed at 4.11 μm. This result has been confirmed using comparison with single arcs that have identical geometry. From A-SRR1 onwards, the right-hand arc is shortened progressively by varying θ. As the angular gaps between the arcs are varied, the two separate resonances of the left-hand and right-hand arcs are modified by the presence of the other arc. In the case of A-SRR1, the asymmetry in length between the left-hand and right-hand arc is still small and a

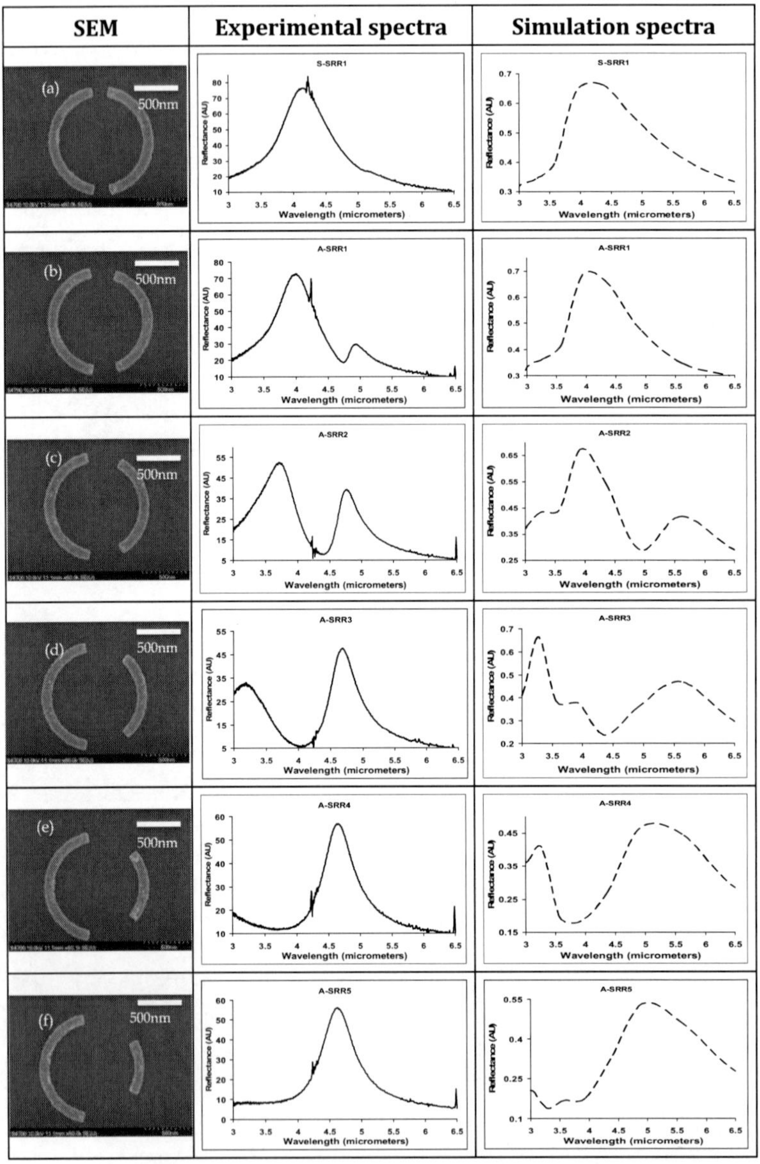

Figure 3.10 The reflectance spectra of the different SRR arrays. The first column shows SEM images of (a) S-SRR1 (θ~10°) (b) A-SRR1 (θ~15°) (c) A-SRR2 (θ~25°) (d) A-SRR3 (θ~35°) (e) A-SRR4 (θ~45°) (f) A-SRR5 (θ~55°) structures. The second column of figures, with solid curves, shows the measured experimental spectra, and the third column of figures, with dashed curves, shows corresponding simulations.

steep edge emerges at 4.7 μm, separating the resonance, at 3.9 μm of the shorter (right-hand) arc from the resonance at 4.9 μm for the longer (left-hand) arc. (Between the two resonances, the reflection dip can be identified as a trapped mode [19], as discussed later). In the cases of A-SRR2 and A-SRR3, as the asymmetry between the lengths of the left-hand and right-hand arcs is varied further, the individual resonances of the two arcs become more distinct, but due to the proximity of the two arcs, the resonances observe are shifted with respect to those exhibited by single arcs alone. This result has been confirmed by simulations of separate arcs in which the individual peaks are closer together. A-SRR2 is marked by the steep edges that emerge in the reflection spectrum between the individual resonances of the left-hand and right-hand arcs, with an increase in the quality factor from 4.7 for the symmetric case S-SRRs in Fig. 3.10a to 9.4 for the longer wavelength reflectance peak of A-SRR2 in Fig. 3.10c. In the corresponding SRRs and A-SRRs with dimensions that lead to resonances in the microwave spectral region [19], at normal incidence the electric field of the linearly polarized plane-wave excitation can excite a circulating current that has the same direction in each arc, and, therefore, coupling into and out of such structures is inherently weak and is zero, in principle, for the exactly symmetric case. In the case of asymmetry between the lengths of the two arcs, the scattered electromagnetic radiation produced by the circulating current configuration is weak but finite, reducing the coupling of the resonant mode to free space and reducing the radiation losses. This behaviour leads to the so-called trapped mode identified by Fedetov *et al.* [19]. When analyzing simulations of the transmission and reflection spectra, a substantial increase in the resonant absorption is observed at the minimum of the reflection dip corresponding to the trapped mode and is consistent with energy being stored and dissipated in the coupled arcs.

For A-SRR3, there is a further shift in the resonance of the right-hand arc to shorter wavelengths, due to further reduction in its length. In the case of A-SRR4, the gap is further increased by additional shortening of the right-hand arc, so the impact of the loading of the non-resonant arc section on the resonance becomes small. Similar behaviour is seen in the case of A-SRR5, where the angular gap is increased even further, resulting in the disappearance of the resonance produced by the shorter (right-hand) arc from the measured spectral range, as it increasingly moves towards shorter wavelengths. It can also be seen that, since the length of the left-hand arc is always kept constant, the resonance position of the longer (left hand) arc does not vary appreciably (moving only from 4.9 μm to 4.6 μm). However, as the right-hand arc is shortened, it is found that the position of its resonance progressively moves towards shorter wavelengths (from 3.9 to 2.5 μm), changing the spectra of the coupled resonance. It can, therefore, be inferred that variation of the angular gap between the two arcs not only increases the quality factor of the resonant peaks in the overall response but also tunes the position of the resonances. The spike present at a wavelength of 4.2 μm in all the experimental spectra indicates the presence of atmospheric carbon dioxide. The Q-factor of the reflection dip of the trapped mode resonance is estimated at a value of 7, here.

The field plots in Fig. 3.11 show the concentration of the electric field distribution towards the ends of the arcs at a wavelength of 3.9 μm for the smaller arc and at a wavelength of 5.6 μm for the larger arc. These wavelengths correspond to the respective reflection peaks. The central plot, at a wavelength of 4.6 μm, corresponds to the reflectance dip of the trapped mode [19], for which the reflectance is sharply reduced and the field distribution in the two arcs is similar.

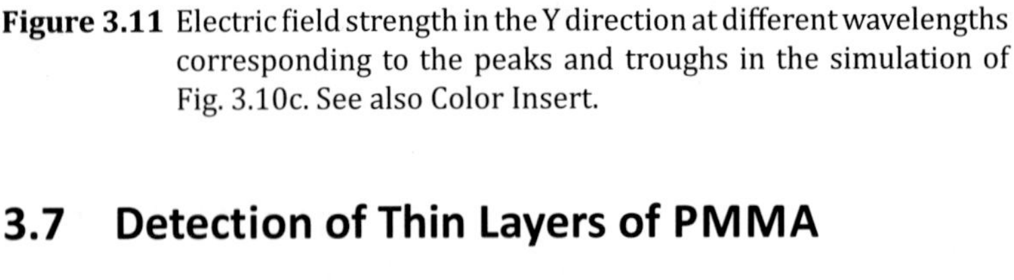

Figure 3.11 Electric field strength in the Y direction at different wavelengths corresponding to the peaks and troughs in the simulation of Fig. 3.10c. See also Color Insert.

3.7 Detection of Thin Layers of PMMA

Many organic compounds contain molecular bonds that exhibit spectral resonances in the mid-infrared range (typically in the 3–5 μm wavelength range) [12]. This property provides the main motivation, in our work, for tuning the resonant response of the A-SRRs to mid-infrared wavelengths. In order to test the sensitivity of the A-SRRs, the fabricated structures were coated with thin films of PMMA, a commonly used positive-tone electron-beam resist. The choice of

PMMA in the present work was made because of the accurate control of the thickness obtainable via control of the spin speed used when coating the sample. We have observed a strong resonant feature at a wavelength of around 5.8 μm due to absorption by the carbonyl bond of PMMA [23]. The use of PMMA as probe material for modifying the plasmonic behaviour has been described in reference [24]. Similar observation of enhanced infrared detection of another organic molecule (ODT) is described in reference [25]. For our experiments, different concentrations of PMMA with molecular weights in the 80–100 k range and around 350 k, dissolved in o-xylene, were spun onto the A-SRR array at different speeds to give uniform thicknesses over the frequency-selective surface (FSS) pattern. The PMMA thickness was varied from 215 nm down to 30 nm, in order to test how effectively the A-SRRs were able to detect its presence. Figure 3.12 shows the shift (experimental and simulation) in the overall

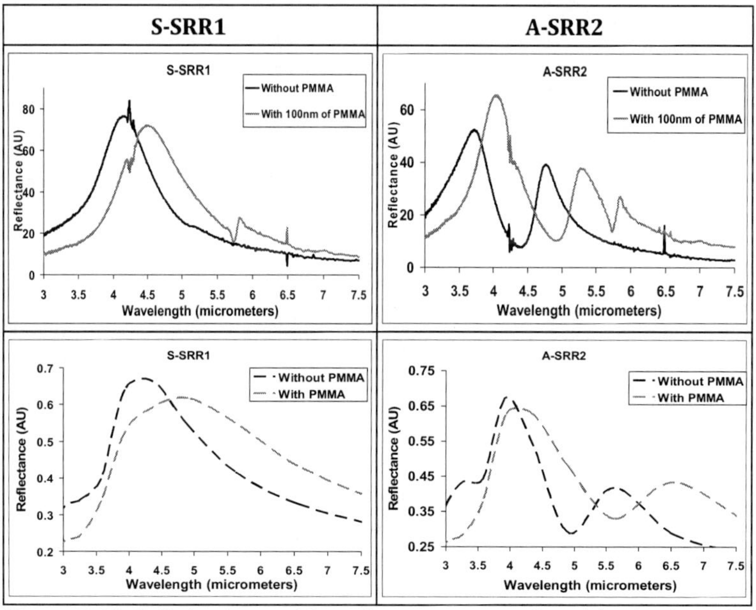

Figure 3.12 Shift in the position of resonance produced by loading the A-SRR array with a 110 nm thick layer of PMMA. The first row with solid lines depicts the experimental results. The second row, with dashed lines in the figures, depicts the corresponding simulations. The black curves show the original resonance without PMMA, and the red curves show the resonance with 110 nm of PMMA deposited on top.

resonant response of the A-SRRs obtained by loading the FSS pattern with 110 nm of PMMA. In the case of the S-SRR1 array, the overall shift due to the loading by the PMMA was less (typically around 0.34 μm in the case of a 100 nm thick PMMA layer). In the case of A-SRR2, where the two distinct resonance peaks become clearly distinct, the shift observed in the position of the resonance was larger (typically around 0.51 μm for a 110 nm thick PMMA layer). Figure 3.13 shows the overall shift in the position of the resonances as a function of the thickness of the PMMA layer. Figure 3.13 shows that A-SRR2 can register the presence of very thin layers of PMMA, as low as ~30 nm, which produces a shift in resonance of 120 nm. To act as an effective biosensor, it will be necessary firstly to attach a vector to the rings that would act as a specific probe for a DNA strand. Thirty nanometres corresponds to a polynucleotide length of around 100 base pair units.

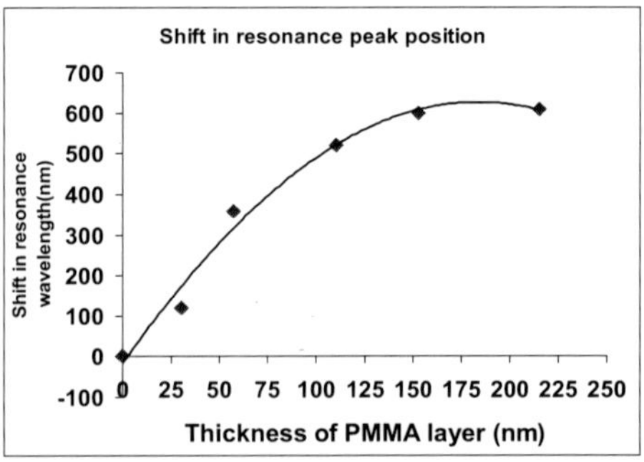

Figure 3.13 The shift in the position of the A-SRR2 to longer wavelength resonance as a function of the thickness of the PMMA layer (solid line is a guide to the eye).

As the thickness of the PMMA increases, the shift in the peak position saturates, which can be attributed to the penetration of the propagating plasmonic mode beyond the PMMA layer becoming negligible. The sensitivity (s) of the A-SRRs towards PMMA is given as [26]

$$s = \frac{\Delta\lambda}{\Delta n} \text{ nm/RIU}$$

where $\Delta\lambda$ is the shift in resonance wavelength, and Δn is the change in the overall refractive index. The unit of the sensitivity (s) is given as nanometres (nm) per Refractive Index Unit (RIU).

From Fig. 3.13, it can be seen that a 215 nm thick PMMA layer shifts the resonance wavelength by 605 nm. A 215 nm thick PMMA layer covers all of the 50 nm thick A-SRR and much above it. Assuming that it replaces all the air above the A-SRR (i.e., the 215 nm thick PMMA layer acts effectively as of infinite thickness above the A-SRRs), the change in refractive index is obtained as 1.49–1.

Therefore, the sensitivity of the A-SRR for 215 nm thick PMMA becomes

$$s = \frac{605}{1.49-1} = 1234.69\ \text{nm/RIU}$$

Thus, it is seen that the sensitivity of A-SRRs for ~200 nm thick PMMA layer is 1234.69 nm/RIU which is a factor of 2.5 times more sensitive from its nearest equivalent photonic crystal resonator utilizing similar techniques with ethanol, as described in reference [26]. In order to detect the presence of small amounts of specific material, e.g., a particular DNA fragment [20–21], complementary receptors would need to be attached to the A-SRRs and, ideally, located near the ends of the arcs, where the electric field strength is greatest. By interpolation of the data in Fig. 3.13, it can be seen that a 5 nm uniform layer of PMMA would still produce a readily observable shift in the resonance peak position of around 25 nm and, if the layer were localized near the ends of the arcs, the sensitivity should be significantly increased further.

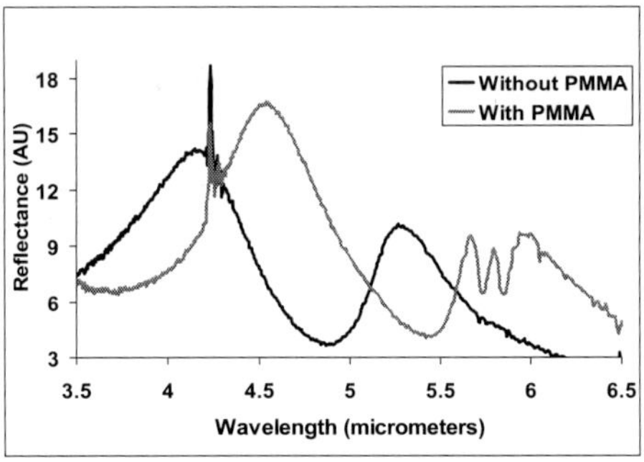

Figure 3.14 The increase in sensitivity of PMMA obtained by matching the resonances. The black curve shows the original resonance without the PMMA, and the red curve shows the resonance with 110 nm of PMMA deposited on top. See also Color Insert.

By changing the overall size of ASRR-2 from 1.2 to 1.35 μm diameter, it is possible to match the longer wavelength resonance of the arcs with the molecular resonance due to the carbonyl group of PMMA. An example of the enhancement in the sensitivity obtained by matching the resonance peak of the A-SRRs to a specific molecular spectral feature is shown in Fig. 3.14. In this case, the molecular resonance at 5.8 μm is due to the presence of the carbonyl group of PMMA.

3.8 Summary and Conclusions

As has been shown throughout this chapter, SRR based metamaterials display very interesting electromagnetic properties. In the first part of the chapter, the fundamental workings and properties of nanoscale SRRs resonating in the (near and mid) infrared and the visible spectrum were presented with the effect of various metals and substrates considered. In the last part of the chapter, the use of steep resonance response of asymmetric split ring resonator (A-SRRs) was described for the optical detection of very small amounts of organic materials (PMMA). Therefore, it can be concluded that metamaterials offer exciting and challenging possibilities for novel electromagnetic responses, and, hence, researchers should continue to utilize their properties for various applications ranging from optical biosensing to telecommunications and more.

References

Journal references

1. Veselago, V.G. (1968). The electrodynamics of substances with simultaneously negative values of ε and μ, *Sov. Phys.*, **10**, pp. 509–514.
2. Pendry, J.B., Holden, A.J., Robbins, D.J., and Stewart, W.J. (1999). Magnetism from conductors and enhanced Non-Linear Phenomena, *IEEE Trans. Microwave Theory Tech.*, **47**, pp. 2075–2084.
3. Pendry, J.B. (2000). Negative refraction makes a perfect lens, *Phys. Rev. Lett.*, **18**, pp. 3966–3969.
4. Linden, S., Enkrich, C., Wegener, M., Zhou, J., Koschny, T., and Soukoulis, C.M. (2004). Magnetic response of metamaterials at 100 THz, *Science*, **306**, pp. 1353–1361.
5. Shalaev, V.M. (2004). Optical negative index metamaterials, *Nat Photon.*, **1**, pp. 41–48.

6. Enkrich, C., Wegener, M., Linden, S., Burger, S., Zschiedrich, L., Schmidt, F., Zhou, J., Koschny, T., and Soukoulis, C.M. (2005). Magnetic metamaterials at telecommunication and visible frequencies, *Phys. Rev. Lett.*, **95**, pp. 203901.

7. Smith, D.R., Schuring, D., and Pendry, J.B. (2002). Negative refraction of modulated electromagnetic waves, *Appl Phys Lett.*, **81**, pp. 2713–2715.

8. Linden, S., Enkrich, C., Dolling, G., Klein, M.W., Zhou, J., Koschny, T., Soukoulis, C.M., Burger, S., Schmidt, F., and Wegener, M. (2006). Photonic metamaterials: Magnetism at optical frequencies, *J. Selected Top. Quantum Electron.*, **12**, pp. 1097–1103.

9. Zhou, J., Koschny, T., Kafesaki, M., Economu, E.N., Pendry, J.B., and Soukoulis, C.M. (2005). Saturation of the magnetic response of split ring resonators at Optical frequencies, *Phys. Rev. Lett.*, **95**, pp. 223902.

10. Klein, M.W., Enkrich, C. , Wegener, M., Soukoulis, C.M., and Linden, S. (2006). Single split ring resonators at optical frequencies: Limits of size scaling, *Opt. Lett.*, **31**. pp. 1259–1261.

11. Camargo, E.A., Chong, H.M.H., and De La Rue, R.M. (2004). 2D photonic crystal thermo-optic switch based on AlGaAs/GaAs epitaxial structure, *Opt. Express*, **12**, pp. 588–592.

12. http://www.rsoftdesign.com/products.php?sub=Component+Design &itm=FullWAVE. 2 December 2009.

13. Zhou, J., Koschny, T., Kafesaki, M., Economu, E.N., Pendry, J.B., and Soukoulis, C.M. (2005). Saturation of the magnetic response of split ring resonators at Optical frequencies, *Phys. Rev. Lett.*, **95**, 223902.

14. Klein, M.W., Enkrich, C., Wegener, M., Soukoulis, C.M., and Linden, S. (2006). Single split ring resonators at optical frequencies: Limits of size scaling, *Opt. Lett.*, **31**, pp. 1259–1261.

15. Tretyakov, S. (2007). On geometrical scaling of split-ring and double bar resonators at optical frequencies, *Metamaterials*, **1**, pp. 40–43.

16. Sheng, Z., and Varadan, V.V. (2007). Tuning the effective properties of metamaterials by changing the substrate properties, *J. Appl. Phys.*, **101**, pp. 014909.

17. Johnson, N.P., Khokhar, A.Z., Chong, H.M.H., De La Rue, R.M., and McMeekin, S.G. (2006). Characterisation at infrared wavelengths of metamaterials formed by thin-film metallic split-ring resonator arrays on silicon. *Electron. Lett.*, **42**, pp. 1117–1119.

18. El-Kady, I., Sigalas, M.M., Biswas, R., Ho, K.M., and Soukoulis, C.M. (2000). Metallic photonic crystals at optical wavelengths, *Phys. Rev. B*, **62**, pp. 15 299–15 302.

19. Fedetov, V.A., Rose, M., Prosvirnin, S.L., Papasimakis, N., and Zheludev, N.I. (2007). Sharp trapped-mode resonances in planar metamaterials with a broken structural symmetry, *Phys. Rev. Lett.*, **99**, 147401.

20. Debus, C., and Bolivar, P.H. (2007). Frequency selective surfaces for high sensitivity terahertz sensing, *Appl. Phys. Lett.*, **91**, 184102.

23. Balamurugan, A., Kannan, S., Selvaraj, V., and Rajeswari, S. (2004). Development and spectral characterization of poly(methyl methacrylate)/hydroxyapatite composite for biomedical applications, *Trends Biomater. Artif. Organs*, **18**, pp. 41–45.

24. Aizpurua, J., Taubner, T., Javier Garcia de Abjo, F., Brehm, M., and Hillenbrand, R. (2008). Substrate-enhanced infrared near-field spectroscopy, *Opt. Express*, **16**, pp. 1529–1545.

25. Neubrech, F., Pucci, A., Cornelius, T.W., Karim, S., Garcia-Etxarri, A., and Aizpurua, J. (2008). Resonant plasmonic and vibrational coupling in a tailored nanoantenna for infrared detection, *Phys. Rev. Lett.*, **101**, 157403.

26. Nunes, P.S., Mortensen, N.A., Kutter, J.P., and Mogensen, K.B. (2008). Photonic crystal resonator integrated in a microfluidic system, *Opt. Lett.*, **33**, pp. 1623–1325.

Proceedings reference

21. Debus, C., and Bolivar, P.H. (2008). Terahertz biosensors based on double split ring arrays, *Proc. SPIE.* 6987, 6987(OU), pp. 3–8.

Book reference

22. Prosvirnin, S., and Zouhdi, S. (2003). *Advances in Electro-magnetics of Complex Media and Metamaterials*, 1st Ed. (Kluwer Academic Publishers, The Netherlands).

Chapter 4

Nanofabrication Technologies for Surface Texturing of Gallium Nitride Photonic Crystal Light-Emitting Diodes

Faiz Rahman,[a,*] Ali Z. Khokhar,[b] Keith Parsons,[c] and Ian M. Watson[d]

[a] *Electrospell Ltd., Block 7, Kelvin Campus, West of Scotland Science Park, Glasgow G20 0SP, UK*

[b] *School of Engineering, Optoelectronics Group, University of Glasgow, Rankine Building, Oakfield Avenue, Glasgow G12 8LT, UK*

[c] *MacDermid Autotype Ltd., Grove Road, Wantage OX12 7BZ, UK*

[d] *Institute of Photonics, University of Strathclyde, Wolfson Centre, 106 Rottenrow, Glasgow G4 0NW, UK*

*Faiz.Rahman@electrospell.com

We describe a comparison of nanofabrication technologies for the fabrication of 2D photonic crystal (PhC) structures on GaN/InGaN blue LEDs. Such devices exhibit enhanced brightness and the possibility of controlling the angular emission profile of emitted light. This chapter describes three nano-lithography techniques for patterning PhC structures on the emitting faces of LEDs: direct-write electron beam lithography (EBL), hard stamp nanoimprint lithography (NIL), and soft-stamp NIL with disposable embossing

Vistas in Nanofabrication
Edited by Faiz Rahman

ISBN 978-981-4364-56-0 (Hardcover), 978-981-4364-57-7 (eBook)
www.panstanford.com

masters. In each case, we describe variations on the technique as well as its advantages and disadvantages.

4.1 Introduction

Environmental-friendly technologies for the generation and utilisation of electric power are rapidly becoming very important. This trend is being driven by a host of factors such as the search for more efficient ways of generating power, research into ways of reducing the wastage of electricity, and the realization of the impact of global warming through increasing CO_2 emissions. A large fraction of generated electric power goes into lighting our homes, work places, and the outdoors. Any savings in power consumption brought out by more energy-efficient lighting technologies is, therefore, very significant in reaching energy conservation milestones. Solid-state lighting based on light-emitting diode (LED) technology is widely regarded as an environmental-friendly alternative to more traditional lighting technologies. This is the reason for the extremely rapid development of LED technology at this time. Many academic, government, and industrial laboratories across the world are working on improving their efficiency and lifetime while at the same time bringing down their cost so that they can replace conventional light sources in a variety of applications [1,2]. Nitride semiconductors from the gallium nitride (GaN) family provide the current materials technology for emitters in the blue and green regions of the electromagnetic spectrum. External quantum efficiencies of up to ~60% have been obtained with commercial packaged blue LED lamps [3]. Phosphide semiconductors meanwhile continue to offer the highest performance at visible red wavelengths. LED-based illumination has been taking over in sectors such as traffic lights and automobile lighting and is now also making strides into areas such as car headlights and street lighting. Electronics manufacturers are already utilising LEDs for backlights in LCD televisions and are interested in general purpose room lighting applications for the near future. The opportunities to develop LEDs with tailored angular emission profiles for more specialised applications, including backlights, LCD projectors, and fibre-optic communications, are also significant. Harbers *et al.* give a recent discussion of the potential for LED solutions in the first two application areas [4]. Incorporation of a PhC structure on the emitting face of LEDs is now recognised as one of the most important technologies for adapting LEDs to various application requirements. The fabrication techniques for PhC and photonic quasi-crystal (PhQC) LEDs described in this chapter will have their first commercial impact in applications requiring increased brightness and optimised angular emission profiles as prime considerations.

All inorganic LEDs generate light inside a high-index medium, and this factor limits the efficiencies attained by simple planar device architectures. This feature arises due to total internal reflection of a large fraction of light within the high refractive index GaN epitaxial layers. As a result, much of the light remains confined as trapped modes inside these layers. The numerical extraction efficiency is only ~4% for GaN-based devices coupling directly into air, but can be increased up to ~10% by encapsulation with a transparent polymer [5]. Most commercial high-power GaN LEDs are now encapsulated in dome-shaped transparent polymers that serve to protect the LED chip as well as increase the coupling of light to the outside ambient. Several other techniques have been developed to extract even more of the internally trapped radiation. These include random surface roughening [6], use of corrugated or patterned substrates [7,8], and patterning the LED surface with a PhC structure [9–12]. PhC and PhQCs function by coupling modes trapped in high-index material into modes which can escape to the outside medium. In this context, the possibility of fabricating thin device structures by laser lift-off (LLO) processing of GaN-based epitaxial structures grown on sapphire is very important [13]. This is because such techniques remove sapphire and, thus, modes trapped in sapphire and also maximise the interaction between trapped modes in GaN and the etched PhC structures. The interaction of guided modes with a PhC/PhQC grating can be maximised when the total GaN thickness is reduced to only a few microns [3]. Furthermore, heat dissipation is also greatly improved over conventional devices, as heat transport through sapphire substrate is replaced by much more efficient heat transfer through metallic back contacts that serve as the substrate. LLO was recently used in the fabrication of thin GaN-based PhC LEDs with a record extraction efficiency exceeding 70% for non-encapsulated devices [14].

In addition to improvement of extraction efficiencies, PhC and PhQC patterning allows tailored modifications to the polar radiation characteristics of LEDs from the near-Lambertian pattern obtained using planar non-encapsulated devices. Much attention has been focussed on enhancing emission in directions close to the surface normal, as desirable in étendue-limited applications [3,15]. However, it is also possible to design PhC and PhQC patterns which cause preferential light emission at large angles to the surface normal, as desirable for certain forms of display backlights, specifically direct illumination of flat panels [16]. Here, highly homogeneous illumination of the application plane is critical, and simulations have shown that the presence of radiation lobes at large angles to the normal can be beneficial for ensuring uniform illumination over large areas. PhQC patterns departing from traditional crystallographic symmetry offer major advantages over their regular PhC counterparts by virtue of the azimuthal uniformity of emission [3,16,17].

All implementations of PhC and PhQC patterns in visible LEDs require a sub-micron patterning capability, and a brief overview of techniques available is included in reference [3]. In this chapter, we report on direct-write EBL and two different embodiments of NIL. NIL techniques have superior potential for mass production and, thus, are attracting considerable attention. Watts *et al.* have reviewed currently available fabrication equipment and economic considerations for commercial implementation of NIL [18]. In all process flows, the PhC/PhQC pattern is first transferred into a mask layer and then transferred into the semiconductor material by a dry-etch process, allowing good control over the hole depth and sidewall profile. Currently, most PhC and PhQC designs assume vertical sidewalls, and etch processes must, therefore, offer a high degree of anisotropy to approximate this idealised geometry in the real device. For GaN-based devices, conventional and thin LEDs made by LLO have different constraints on hole depth, as can be understood with reference to the typical p-i-n device structure on sapphire shown in Fig. 4.1. The topmost p-type layers of commercial LED wafers above the active region are normally kept to a thickness of ~150 nm. Therefore, in devices with the conventional (so-called p-up) geometry, the PhC/PhQC hole etch depth cannot exceed this value without catastrophic shorting of the junction and/or etch damage to the active region. In contrast, in thin GaN devices made by LLO processing, the depth of holes etched into n-type GaN can be significantly greater, as illustrated by published device results [14, 19].

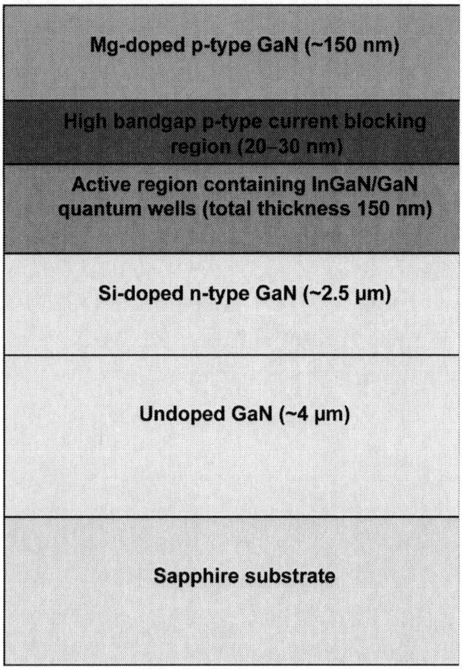

Figure 4.1 Structure of a typical epitaxial GaN blue LED wafer material showing the positions and thicknesses of various layers.

This chapter presents results from an integrated programme of development and optimisation of patterning and dry-etch nanofabrication technologies for GaN-based LEDs. We show results on the processing of as-grown GaN LED wafers on sapphire, corresponding to process flows for conventional p-up LEDs. The processing techniques developed are applicable to thin GaN devices made by LLO routes, as well as to LEDs fabricated from non-nitride materials. The results to date have been obtained using 2 inch sapphire based wafers. However, the techniques developed are applicable to GaN device structures on larger sapphire wafers [20] and lower cost large silicon substrates [21]. Both these approaches are being actively pursued to reduce the epitaxy-related cost component of GaN LEDs. Our specific motivation concerned the development of PhQC LEDs for direct flat panel backlighting for LCDs. Results on the performance characteristics of devices made by optimised fabrication routes have been reported elsewhere [22].

4.2 PhC Nanofabrication with Direct-Write e-Beam Lithography

Direct-write e-beam lithography has been used for many years to write nanometre-scale features on semiconductors. PhC structures consisting of arrays of holes, several hundred nanometres in diameter, and spaced a few hundred nanometres apart, can be, and are, easily written by e-beam lithography. This technique, though not suitable for commercial production, is very well suited to R&D operations as no masks or stamps need to be prepared. We have carried out PhC patterning using the direct-write technique with a view to exploring the suitability of various resists for defining these structures.

In this work, we patterned a 12-fold PhQC design [23] with different hole diameters and pitches on GaN LED epi-layer wafers. It has been proved that by changing the hole diameter and/or pitch the emission profile of PhC LEDs could be modified [24]. Various e-beam resists such as Hydrogen SilsesQuioxane (HSQ), a silica-based resist, NEB31, a chemically-amplified resist from Sumitomo, ZEP-520A, a non chemically-amplified resist from Nippon Zeon, and PolyMethyl MethAcrylate (PMMA) were used to write PhC patterns. HSQ and NEB31 are negative-tone resists whereas ZEP-520A and PMMA are positive-tone resists. Leica VB6 e-beam tool was used to write patterns on commercial 2" GaN LED epi-layer wafers, sourced from Lumi-innovations. VB6 is a very high resolution machine and can produce e-beam spot sizes down to around 4 nm diameter on a design grid as small as 1.25 nm.

GaN LED epi-layer wafers were cleaned with acetone, methanol, isopropanol, and purified water before spinning on resist. All GaN samples were coated with 200 nm thick resist. It was found that 30 nm of an Aluminum (Al) layer on top of the resist mask gives very high quality patterns; Al layer acts as a discharge layer for e-beam lithography. However, sometimes Al may be difficult to remove in which case conducting organic polymer based charge dissipation layers can be utilised [25, 26]. These materials can be applied easily using a spin-coating process and can be removed by simply washing away in water.

Our work revealed that HSQ has a number of advantages over other e-beam resists when it comes to patterning GaN LED wafers, such as

1. HSQ acts as a direct hard mask for dry etching as the cured material is a fairly dense silica layer. For this reason, while using HSQ there is no need to use a separate metal hard mask. We found no significant change in the 200 nm HSQ mask during the etching of up to 300 nm of GaN LED epi-layer wafer. With other e-beam resists, such as ZEP-520A or PMMA, we needed to use a metal hard mask such as a NiCr mask on top of the e-beam resist [27].
2. HSQ is not sensitive to white light. We deposited 30 nm of Al layer on the top of resist to act as a discharge layer for e-beam writing. The Al layer was deposited by using a Plassys e-beam evaporator tool. It was found that NEB31 resist was exposed during a similar deposition of Al layer.
3. HSQ can be easily removed from GaN LED epi-layer wafers using buffered HF but care has to be taken to keep this step as short as possible.

With HSQ, the best e-beam dose was found to be 700–800 μC/cm^2 at 100KV e-beam accelerating voltage. This was determined by using a dose estimation matrix. The hole size can be varied by changing the e-beam dose [28]. The criterion chosen for the best e-beam dose was the achievement of exactly the same size hole diameter in the HSQ resist as was provided in simulated designs. Each sample contained an array of 4 × 4 patterns together with four alignment markers. These alignment markers were used to fabricate *n* and *p* contact after dry-etching of the PhC pattern. Each pattern had dimensions of 300 μm × 300 μm. Samples also contained blank devices. These blank devices were intended to compare LEDs with and without PhQC pattern. After e-beam writing, the samples were developed in Tetra Methyl Ammonium Hydroxide (TMAH) resist developer and inspected under an optical microscope. Patterns in HSQ were transferred to the epi-layer by a chlorine-based dry-etch process that is described in Section 4.5. After dry-etching some samples were cleaved and their plan and cross-sectional views obtained with an SEM. Some representative micrographs appear in Fig. 4.2 here.

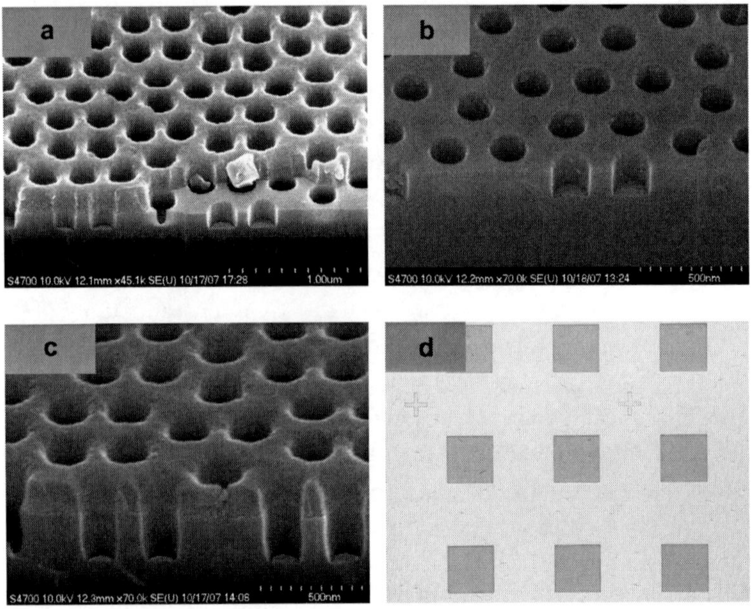

Figure 4.2 (a) SEM micrograph of GaN epi-layer wafer. Arrow indicates interface between HSQ and GaN epi-layer wafer. The etching depth is ~150 nm. (b) SEM micrograph of GaN epi-layer wafer after removing HSQ by using buffered HF. (c) SEM micrograph of GaN epi-layer wafer with HSQ. Arrow indicates interface between HSQ and GaN epi-layer wafer. Etching depth is ~300 nm. Note: No significant change has been observed in HSQ after etching 300 nm GaN epi-layer wafer by using ICP tool. (d) Optical micrograph of GaN epi-layer wafer with 300 μm × 300 μm patterns with alignment markers.

4.3 PhC Nanofabrication with Rigid-Stamp Nanoimprint Lithography

Direct-write e-beam lithography is a serial process and is, thus, far too slow for use as a viable commercial patterning process. Patterning even a full 2 inch diameter wafer with PhC patterns using this technique will cost upwards of $10 K which is clearly unacceptable. NIL has emerged in recent years as an affordable technique for the mass replication of sub-micron size structures over large areas. This is also very relevant because efforts are being made to produce large diameter GaN LED wafers in order to bring down the cost of

LED chips. Micro and nano-optics are the areas where NIL has been especially popular, and, thus, commercial PhC patterning has been mostly carried out with this technique. Among the advantages of NIL are relatively low cost of capital equipment and high production speed which combine to make NIL patterning very affordable [29]. We have explored the two basic variants of NIL to imprint PhC patterns on GaN LED wafers with a view to understanding yield limiting factors.

In this work, the following two rigid-stamp NIL approaches were used.

1. Heat and Pressure NIL
2. Flash Imprint NIL

4.3.1 Heat and Pressure NIL

A silicon stamp was first fabricated for Heat and Pressure (H&P) NIL. A clean 15 mm × 15 mm Si substrate with a thickness of 500 μm was spin-coated with HSQ. An inverse design of PhQC was transferred to HSQ by using e-beam lithography. The design was further transferred to the Si substrate by Inductively Coupled Plasma (ICP) dry etching [30]. The gas mixture used for dry etching was SF_6/C_4F_8. The dry-etching depth was found to be 260 nm as measured with an SEM.

This work used a commercial nanoimprint tool called nanoIMPRINT from Obducat in Sweden. The commercially available NIL polymer mr-I-9020 [31] was spin-coated on a GaN epi-layer wafer at 2000 RPM. The Si stamp was placed face down on the GaN epi-layer wafer. The entire assembly was then placed in the nano-imprinter. The glass transition temperate (T_g) of mr-I-9020 is above 100°C. The temperature was raised above the T_g to 120°C and a pressure of 40 Bar was applied. At high temperatures, the viscosity of the polymer is low enough to fill all the cavities in the stamp. After a few minutes, the polymer started cross-linking, and this caused an increase in viscosity. The polymer solidified and the shape formation of the imprint finished after 15 minutes. The stamp was detached from the GaN LED epi-layer wafer at a temperature below the T_g of mr-I-9020. During the H&P NIL process, pressure was used to keep the Si stamp attached with the substrate during the stamp cavity filling process.

After imprinting, the GaN LED epi-layer wafer was examined under SEM. Very encouraging results were achieved with H&P NIL, as seen in Fig. 4.3. It was found that this NIL technique could be used to imprint on smooth as well as on rough surfaces. The disadvantage of H&P NIL technique is that, due to its slow speed, it cannot be used for industrial production and for processes that need alignment of the stamp with the substrate.

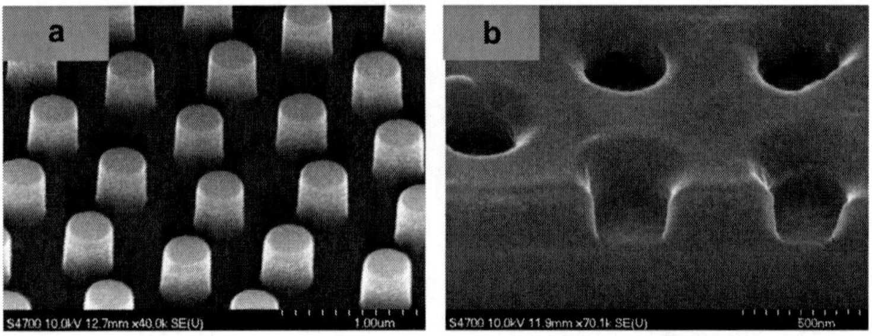

Figure 4.3 (a) SEM micrograph of silicon stamp after dry-etching by using ICP tool. HSQ was removed by using buffered HF. (b) SEM micrograph of pattern-imprint patterns on GaN epi-layer by using an Obducat nanoIMPRINT imprinter tool.

4.3.2 Flash NIL

Flash NIL requires UV exposure for cross-linking the NIL resist. For this reason, a quartz stamp was prepared for Flash NIL. In this work, a quartz substrate having dimensions of 65 mm × 65 mm and thickness of 6 mm was used. NEB31 negative-tone e-beam resist was spin-coated on a quartz substrate. NEB31 mask has an advantage over HSQ in that its etch rate is lower during quartz etching. An Inverse PhQC pattern with pillars of 550 nm diameter was written on the NEB31 resist layer. A BP80 dry-etch machine and CHF_3 gas were used for dry-etching the quartz substrate. After dry-etching, the patterns were examined under SEM. Vertical etching profile with depth of 180 nm was found in SEM observations.

In this work, the commercially available Flash NIL polymer, mr-UVCUR06 [32], was spin-coated on GaN epi-layer wafer at 3000 RPM. After spinning, the sample was baked at 90°C for 60 sec. Flash NIL was done by using the hard contact mode of an MA6 mask aligner. The sample was exposed to UV for 5 minutes. After Flash NIL, the substrate was cleaved and observed under SEM. It was found that the thickness of the residual layer was 20 nm whereas the NIL feature depth was 180 nm. SEM micrographs are shown in Fig. 4.4.

Figure 4.4 (a) SEM micrograph of 300 μm × 300 μm pattern fabricated on Quartz stamp by using e-beam lithography and dry etching. (b) SEM micrograph of Quartz stamp at high magnification. (c) SEM micrograph of GaN epi-layer wafer after imprinting by using the Flash NIL technique. (d) SEM micrograph of GaN epi-layer wafer after the pattern was transferred into GaN by using the ICP tool.

4.4 PhC Nanofabrication with Flexible Stamp Nanoimprint Lithography

Rigid-stamp NIL has its limitations. Frequently encountered problems include release from the master, stitching errors in the step and repeat process, damage to bowed wafers, especially in high-pressure imprint processes, and clearing the inevitable residual layer when device fabrication involves pattern transfer into the underlying substrate. An alternative wafer-scale nanoimprint process is to use flexible-stamp NIL [33–35]. In principle, this nanoimprint technique reduces or even overcomes the limitations of other methods and lends itself to large scale low cost roll-to-roll processing [36]. The technique exploits the concept of a single use disposable master. It also relies upon the development of imprint resists with low surface energy that allow easy release of the disposable master. Imprint resists can be formulated to have sufficient etch selectivity to be compatible with deep relief pattern transfer into GaN substrates.

Master structures can be made using a variety of techniques, such as photolithography in photoresist or direct-write electron-beam patterning, and then transferred to nickel replicas by standard electroforming techniques. The pattern on the nickel can then be transferred to the surface of a polyethylene terephthalate (PET) film using a roll-to-roll UV replication process. This provides a large supply (Hundreds of metres) of "disposable masters" for nanoimprinting.

The UV cross-linking materials used in the replication process comprise a conventional acrylate system containing a small percentage of fluoroacrylate. The formulation of each resist can be designed so that spin-coated films have a reduced surface energy enabling clean release of the master without additional processing steps. The disposable masters are flexible; their stiffness is determined by the combination of the PET-backing film and the formulation of the lacquer containing the relief.

The first step in the process is to spin-coat the substrate (typically gallium nitride) with a thin film of UV-sensitive imprint resist, followed by a short low temperature pre-bake to evaporate excess solvent and to promote adhesion of the polymer film to the wafer. The imprint step involves using a roller to press the disposable master into the resist with only light force, as shown schematically in Fig. 4.5. Unlike conventional imprinting, the line of contact of the flexible master makes contact pressure virtually irrelevant. In many cases, a simple hand roller is fit for purpose. Experiments show that the degree of filling and the thickness of the residual layer are controlled by the thickness of the spin-coated layer rather than the pressure.

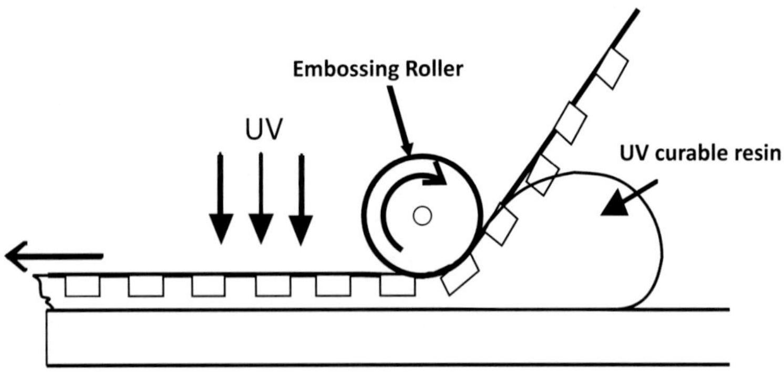

Figure 4.5 Schematic of pattern transfer with a disposable master film. A roller is shown pressing the patterned film into a UV-curable polymer layer on a harder substrate. Photo-curing of the imprinted polymer allows clean separation of the master.

After imprinting, the next step is to cure the resist. This involves a short exposure to UV light and an optional thermal cure to complete the cross-linking of the imprint resist. Finally the disposable master is released by peeling it from the imprinted substrate. It is often found that the disposable master is already partially released after the cure, owing to the low-surface energy design of the imprint resists, eliminating the need for force at a critical stage. Figure 4.6 shows a pattern that has been transferred to a wafer using this process. The pattern consists of an array of 400 nm diameter holes, shown at the right. The left-hand panel in this figure shows the master stamp used in this process. It was directly written onto silicon by EBL.

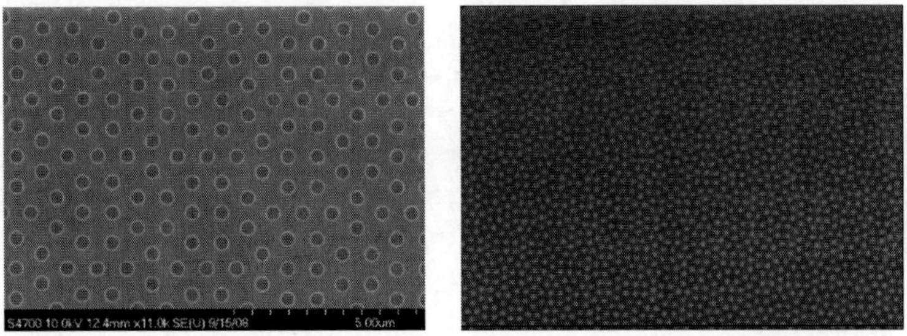

Figure 4.6 Left-hand image shows electron-beam written master. Right-hand image shows patterned imprint resist on wafer after flexible master nanoimprint lithography.

Given that disposable masters are produced on a thin PET film, in a roll-to-roll process, they are highly cost effective to produce. A single rigid master structure can be used to produce many hundreds or thousands of metres of disposable master film. As implied by the description, the individual disposable master films are cheap enough to be discarded after each use. This provides a distinct advantage over rigid stamps, which are more expensive to produce as they are made from more expensive materials (metals, silicones) which also need to be much thicker to provide the required rigidity and robustness.

There are, however, several issues that can limit the applicability of flexible master NIL. Firstly, if the pattern is to be successfully transferred to a GaN wafer by means of etching, then the "residual layer", or layer of imprint resist in between the pattern and the wafer, needs to be as thin as possible. This implies excellent control over imprint-resist spin coating and may impact the choice of materials that can be used in the imprint resist formulation. Secondly, the imprint resist must be formulated to provide the appropriate etch selectivity. The choice of materials available to impart etch resistance may not correspond with the requirements for achieving a thin residual layer. We have successfully explored the use of nano silica-loaded NIL resist in this work for imparting increased etch resistance to patterned resist. Special resist formulations containing an optimum amount of nano silica can significantly reduce resist erosion during dry etching and, thus, provide a higher etch selectivity relative to GaN. This is distinct from the incorporation of oxetanyl xilsesquioxane (OXSQ) units that contain molecular scale Si-O cage units [36]. In our work, we made use of much bigger (~40 nm diameter) nano-silica particles that resulted in etch rates down to 50 nm/min, i.e., about a factor of four lower than the GaN etch rate under the same conditions. Thirdly, the flexible master must be manufactured from thin PET to maintain flexibility and the UV-replicated coating containing the relief must not limit the flexibility. This implies a thin replicated coating and therefore can restrict the use of deeper surface relief profiles.

4.5 Dry Etching for Pattern Transfer

ICP etching is now used in most fabrication of GaN-based LEDs [37]. This process offers high-density plasmas with concentrations of reactive species in the 10^{12} cm^{-3} range and also gives wide scope for manipulating the physical and chemical contributions to the overall etch process. In standard ICP, processing of GaN LED structures grown on insulating sapphire substrates, etching proceeds through the p-type layers and active region to produce mesas. Control of the etch depth and sidewall profile are relatively non-critical in this application. However, in the fabrication of PhC and PhQC LEDs, ICP etching has a second most demanding role in transferring the photonic microstructure from a mask material into the GaN itself. All examples in this chapter involve pattern transfer into the topmost p-type GaN layers of LED structures grown on c-plane sapphire by metal organic vapour phase epitaxy. In this situation, the GaN is etched on the gallium-polar (0001) crystal plane of the wurtzite–phase material. Processing of flip-chip and vertical-current LEDs mentioned in the introduction involves dry etching of GaN on the nitrogen-polar (000$\bar{1}$) crystal plane, which may show dissimilar behaviour with respect to both etch rate and roughening [38, 39]. The need for separate optimisation of such processes has not been widely recognised in the literature.

The ICP etch tool used in work at Strathclyde University is a Surface Technology Systems LPX system, which is equipped with a laser reflectometer. Chlorine-argon gas mixtures were used, and samples were mounted with grease or wax on silicon carrier wafers. For routine etching of LED mesas, ICP etch processes are operated at etch rates often in the 500–1000 nm/min range. However, considerably slower etch rates are desirable for accurate control of PhC and PhQC hole depths. More specifically, the total process time should be long compared to the stabilisation time of a few seconds expected after the plasma is first struck. Etch rates may be controlled effectively by adjusting the inductively-coupled coil power; with zero coil power the process becomes equivalent to a simple capacitively-coupled reactive-ion etch process. Figure 4.7 shows the dependence between etch rate on the GaN (0001) face and coil power, holding other process parameters constant. These included a platen power of 200 W, a pressure of 5 milliTorr, a chlorine flow rate of 30 standard cubic centimetres per minute (sccm), an argon flow rate of 10 sccm, and a chuck set point temperature of 30°C. The recipe with a coil power of 500 W is representative of one used for mesa etching in standard LED processing. The recipe with a 50 W coil power, corresponding to an etch rate in the plot of 210 nm/min, was chosen for the etching of PhQC structures.

HSQ masks patterned by e-beam lithography, as discussed in Section 4.2, proved satisfactory for the transfer of PQC patterns into GaN. The etch selectively between HSQ and GaN is similar to the value of 1:6 obtained between conventional SiO_2 hard masks and GaN. Etch rates measured from profiling of long step features also transferred accurately to sub-micron HSQ patterns. Figure 4.8 shows a representative atomic force microscope (AFM) image of a PhQC test patterns etched to a nominal hole depth of 105 nm. The measured hole depth on this sample, averaged from five locations, was 99 nm. A second sample etched to a target hole depth of 55 nm was also examined by AFM. The measured hole depth, averaged from three locations, was 56 nm.

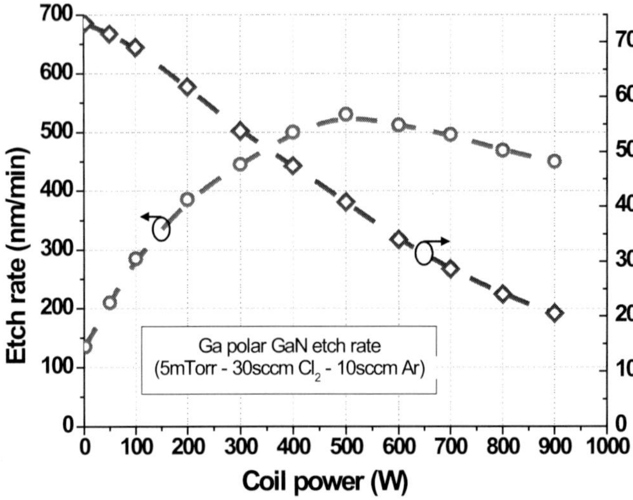

Figure 4.7 Etch rate on the GaN (0001) crystal face and dc self-bias as a function of ICP coil power. Other process parameters were kept constant and are detailed in the text.

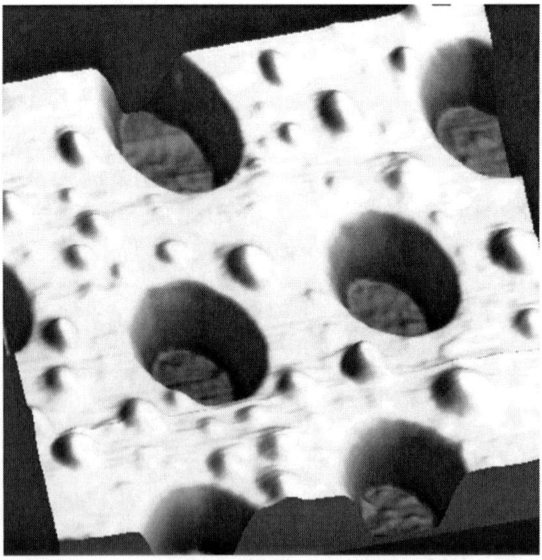

Figure 4.8 Tapping-mode AFM image of 2 μm × 2 μm area of a PQC test pattern etched into GaN, with a hole diameter of 400 nm and nominal hole depth of 105 nm. The protrusions on the surface surrounding the holes are believed to originate during epitaxial growth.

Figure 4.7 also shows the trend in dc self-bias with changing coil power, which is relevant to the sidewall slope obtained in ICP etching. Larger dc biases at lower coil powers enhance the sputter-based physical contribution and thus favour the formation of near-vertical sidewalls, if other factors are favourable. However, the actual sidewall slope obtained is also a function of the HSQ mask profile, cross-sectional SEM views of which are shown in Fig. 4.2. Experimental determination of the sidewall slope of PhQC holes in GaN of depths of interest for working devices (typically <150 nm) is challenging. However, cross-sectional SEM views of a specially prepared test sample with a hole depth of 800 nm showed that the GaN sidewalls are within 10° of vertical with the preferred etch recipe.

As noted already, the ICP etch tool employed features a laser reflectometer allowing real-time in situ monitoring of the etch process through thin film interference effects. This monitoring technique is now very common in epitaxial growth but is also widely adopted for endpoint detection in plasma-etch systems [40]. The monitoring wavelength in this study was 670 nm, and the beam was focussed onto the sample at normal incidence through a window in the ceiling of the process chamber. The refractive index difference of ~0.6 between GaN and sapphire substrates produces large changes in sample reflectance as the optical thickness of the GaN layer changes. Derivation of GaN etch rates is, thus, very straightforward and is analogous to the already widespread use of reflectometry in growth monitoring [41]. Provided the GaN layer thickness is sufficiently large, etch rates under several different etch conditions can be measured in a single process run. This methodology was used to obtain the data plotted in Fig. 4.7, using a uniform GaN epilayer of nominal 4 μm thickness. The ternary alloy layers of LED structures, as illustrated in Fig. 4.1, do not greatly complicate reflectance monitoring per se, since their refractive indices do not differ from that of GaN by more than a very few percent. However, etch rates of aluminium-containing alloys are lower than those of GaN, and thus simple GaN epilayers are preferred to full LED structures for studies of GaN etch rates.

Reflectance monitoring also allows etch rates of novel mask materials and the selectivities with respect to GaN etch rates, to be determined in a straightforward fashion. These measurements are particularly important for polymeric nanoimprint resists, where the finite thickness of the residual layer needs to be taken into account for pattern-transfer into the underlying GaN. The polymer formulations themselves have high transparency in the visible, and a suitable sample type to study the etch resistance is an unpatterned layer of the polymer on an opaque silicon substrate. Illustrative results are shown in Fig. 4.9, which is a plot of reflectance versus time during exposure of a 230 nm layer of a cured acrylate-fluoroacrylate resin, as discussed in Section 4.4, to the etch conditions used for pattern-transfer into GaN. Full removal of the polymer is evidenced by the constant reflectance value reached after an etch time of ~25 sec. The flat reflectance signal in the first 2 sec after the process-start corresponds to a "dead time" caused by plasma stabilisation. The effective etch time of 23 sec for the polymer layer, thus, corresponds to an average etch rate of 600 nm/min, ~3× that of GaN under the same conditions. Even this relatively low etch resistance proved sufficient for the transfer of sub-micron motheye-type patterns [42], produced by the flexible-stamp NIL method discussed in Section 4.4, into GaN layers. However, much lower polymer etch rates down to 50 nm/min have been demonstrated using modified resins employed in the most recent work on PQC-LED fabrication. That work shows that using proprietary blends of acrylate-fluoroacrylate resin, the ICP etch rate of the resist can be reduced from 3× that of GaN to less than that of GaN. While resist without nano silica can be successfully used for the transfer of shallow patterns to GaN, silica-loaded resists can enable the transfer of significantly deeper patterns.

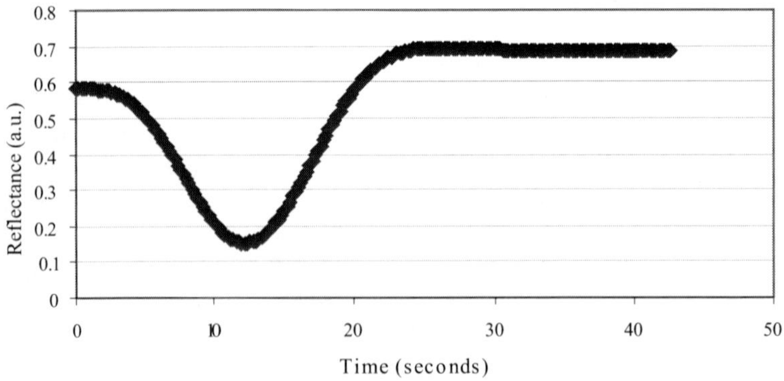

Figure 4.9 Plot of reflectance versus time measured during a test of the etch resistance of an acrylate-based nanoimprint resin to the ICP etch recipe used for pattern transfer into GaN.

After pattern-transfer, LED structures measuring 350 μm × 350 μm were fabricated using standard photolithography, dry etching, and metallization techniques. An SEM micrograph of our finished PhQC LED device appears in Fig. 4.10.

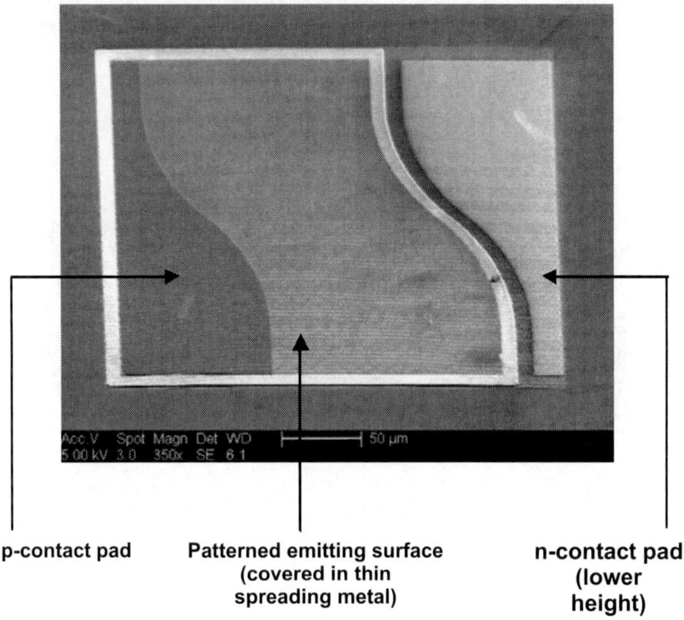

Figure 4.10 SEM micrograph of a finished PhQC LED device showing the central area patterned with PhQC structure as well as the *n* and *p* contacts to the right and left, respectively.

The *n*-contact pad, seen to the right, required etching down to the *n* type contact layer of the epitaxial structure whereas the p-contact pad was formed directly on top of the wafer. A thin Ni/Au bilayer spreading metal was deposited on the top face of the LED to distribute current evenly across the entire cross section of the device. This transparent conductive layer is needed in all large area LEDs because the relatively low electrical conductivity of *p* type GaN would otherwise lead to current crowding effects resulting in uneven illumination from the emitting face of the LED. In order to further reduce current crowding effects and resulting non-uniform thermal effects in the LED chip, we utilised anode and cathode electrode designs developed by Bulashevich and colleagues from simulation of such effects in blue GaN LEDs [43]. The curved contours of this design are clearly seen in Fig. 4.10. We tested our devices both on wafer and after dicing and bonding on to TO-style headers. Their I-V characteristics are essentially identical to unpatterned reference devices made on the same chips, thus showing that the PhQC patterning process doesn't cause electrical degradation of LEDs. These devices performed very well in our tests, generating strong and uniform emission from the top emitting face.

4.6 Summary

We have described nanofabrication technologies for patterning PhC structures on LED wafers. These include direct-write EBL, hard stamp NIL, soft-stamp nanoimprint lithography for patterning resist, and inductively-coupled plasma-based dry etching for transferring patterns into semiconductor material. Process details have been provided which clearly show that EBL is desirable for R&D purposes whereas nanoimprint lithography is the only option for large scale device manufacturing. Soft-stamp nanoimprint lithography with disposable masters, in particular, appears to be the lowest cost technique for the mass manufacture of PhC LEDs. ICP dry etching is also demonstrated as ideal for transferring patterns defined in resist to the underlying semiconductor wafer material with high fidelity.

References

1. G. Held, *Introduction to Light Emitting Diode Technology and Applications*, Auerbach Publications, Boca Raton, 2008, ISBN: 1420076620.
2. A. Zukauskas, M.S. Shur, and R. Caska, *Introduction to Solid-State Lighting*, John Wiley and Sons Inc., New York, 2002, ISBN: 0471215740.
3. M.R. Krames, O.B. Shchekin, R. Mueller-Mach, G.O. Mueller, L. Zhou, and M. George Craford, *J. Display Technol.*, **3** (2007) 160–174.
4. G. Harbers, S.J. Bierhuizen, and M.R. Krames, *J. Display Technol.*, **3** (2007) 98–108.
5. C. Wiesmann, K. Bergenek, N. Linder, and U.T. Schwarz, *Laser Photon Rev.*, **3** (2009) 262–286.
6. Chul Huh, Kug-Seung Lee, Eun-Jeong Kang, and Seong-Ju Park, *J. Appl. Phys.*, **93** (2003) 9383–9386.
7. T. Mukai, S. Nagahama, T. Yamamoto, M. Sano, D. Morita, M. Yamamoto, M. Nonaka, K. Yasutomo, and K. Akaishi, *Phys. Stat. Sol.*, **(c) 2** (2005) 3884–3886.

8. K. Tadatomo, H. Okagawa, Y. Ohuchi, T. Tsunekawa, Y. Imada, M. Kato, and T. Taguchi, *Jpn. J. Appl. Phys.*, **40** (2001) L583.
9. T.N. Oder, J. Shakya, J.Y. Lin, and H.X. Jiang, *Appl. Phys. Lett.*, **83** (2003) 1231.
10. H.K. Cho, J. Jang, J.-H. Choi, J. Choi, J. Kim, J.S. Lee, B. Lee, Y.H. Choe, K.-D. Lee, S.H. Kim, K. Lee, S.-K. Kim, and Y.-H. Lee, *Opt.* Express, **14** (2006) 8654–8660.
11. A. David, T. Fujii, R. Sharma, K. McGroddy, S. Nakamura, S.P. DenBaars, E.L. Hu, C. Weisbuch, and H. Benisty, *Appl. Phys. Lett.*, **88** (2006) 0611241–0611243.
12. F. Rahman, and R. De La Rue, *Photon Spectra*, **41** (2007) 52–55.
13. P.R. Tavernier, and D.R. Clarke, *J. Appl. Phys.*, **89** (2001) 1527–1536.
14. J.J. Wierer Jr, A. David, and M.M. Megens, *Nat. Photon.*, **3** (2009) 163–169.
15. K. McGroddy, A. David, E. Matioli, M. Iza, S. Nakamura, S. DenBaars, J.S. Speck, C. Weisbuch, and E.L. Hu, *Appl. Phys. Lett.*, **93** (2008) 1035021–1035023.
16. M.D.B. Charlton, M.E. Zoorob, and T. Lee, *Proc. SPIE*, **6486** (2007) 64860R1-10.
17. A. David, H. Benisty, and C. Weisbuch, *J. Display Technol.*, **3** (2007) 133–148.
18. M.P.C. Watts, M. Zoorob, T. Lee, and J. McKenzie, *Proc. SPIE*, **6462** (2007) 64620N1-12.
19. H.W. Huang, C.H. Lin, K.Y. Lee, C.C. Yu, J.K. Huang, B.D. Lee, H.C. Kuo, K.M. Leung, and S.C. Wang, *Semicond. Sci. Technol.*, **24** (2009) 085008-1-5.
20. J. Soellner, O. Schoen, A. Alam, B. Schineller, J. Kaeppeler, and M. Heuken, *Thin Solid Films*, **515** (2007) 4372–4364.
21. J. Li, J.Y. Lin, and H.X. Jiang, *Appl. Phys. Lett.*, **88** (2006) 171909-1-3.
22. A.Z. Khokhar, K. Parsons, G. Hubbard, I.M. Watson, F. Rahman, D.S. Macintyre, C. Xiong, D. Massoubre, Z. Gong, E. Gu, N.P. Johnson, R.M. De La Rue, M.D. Dawson, S.J. Abbott, M.D.B. Charlton, and M. Tillin, *Appl. Opt.*, **50** (2011) 3233–3239.
23. M.E. Zoorob, M.D.B. Charlton, G.J. Parker, J.J. Baumberg, and M. Netti, *Nature*, **404** (2000) 740–743.
24. K. Kim, J. Choi, J.B. Park, S.C. Jeon, J.S. Kim, and H.M. Lee, *IEEE Photon. Technol. Lett.*, **20** (2008) 1455–1457.

25. R. Dylewicz, S. Lis, R.M. De La Rue, and F. Rahman, *J. Vacuum Sci. Technol. B*, **28** (2010) 817–822.

26. R. Dylewicz, S. Lis, R.M. De La Rue, and F. Rahman, *IEE Electron. Lett.*, **46** (2010) 1025–1027.

27. B. Rong , R. Cheung, Gao, and M. Kamp, *Microelectron. Eng.*, **53** (2000) 419–422.

28. P.A. Shields, M.D.B. Charlton, T. Lee, M.E. Zoorob, D.W.E. Allsopp, and W.N. Wang, *IEEE J. Sel. Top. Quantum Electron.*, **15** (2009) 1269–1274.

29. S.H. Kim, K.-D. Lee, J.-Y. Kim, M.-K. Kwon, and S.-J. Park, *Nanotechnology*, **18** (2007) 055306–055311.

30. V.A. Yunkin, D. Fischer, and E. Voges, *Microelectron. Eng.*, **27** (1995) 463–466.

31. K. Pfeiffer, F. Reuther, M. Fink, G. Gruetzner, P. Carlberg, I. Maximov, L. Montelius , J. Seekamp, S. Zankovych, M. Sotomayor-Torres, H. Schulz, and H.-C. Scheer, *Microelectron. Eng.*, **67–68** (2003) 266–273.

32. M. Vogler, S. Wiedenberg , M. Muhlberger , I. Bergmair ,T. Glinsner , H. Schmidt, E.-B. Kley , and G. Grutzner, *Microelectron. Eng.*, **84** (2007) 984–988.

33. L.J. Guo, *Adv. Mater.*, **19** (2007) 495–513.

34. J.J. Lee, S.Y. Park, K.B. Choi, and G.H. Kim, *Microelectron. Eng.*, **85** (2008) 861–865.

35. S.H. Ahn, and L.J. Guo, *Adv. Mater.*, **20** (2008) 2044–2049.

36. G. Hubbard, S.J. Abbott, Q. Chen, D.W.E. Allsopp, W.N. Wang, C.R. Bowen, R. Stevens, A. Satka, D. Hasko, F. Uherek, and J. Kovac, *Phys. E*, **41** (2009) 1118–1121.

37. K.H. Baik, and S.J. Pearton, *Appl. Surf. Sci.*, **255** (2009) 5948–5951.

38. F. Rizzi, E. Gu, M.D. Dawson, I.M. Watson, R.W. Martin, X.N. Kang, and G.Y. Zhang, *J. Vacuum Sci. Technol. A*, **25** (2007) 252–260.

39. I. Waki, M. Iza, J.S. Speck, S.P. DenBaars, and S. Nakmuara, *J. Appl. Phys.*, **45** (2006) 720–723, in Japanese.

40. I.P. Herman, *Ann. Rev. Phys. Chem.*, **54** (2003) 277–305.

41. C. Liu, and I.M. Watson, *Semicond. Sci. Technol.*, **22** (2007) 629–635.

42. H. Kasugai, Y. Miyake, A. Honshio, S. Mishima, T. Kawashima, K. Iida, M. Iwaya, S. Kamiyama, H. Amano, I. Akasaki, H. Kinoshita, and H. Shiomi, *Jpn. J. Appl. Phys.*, **44** (2005) 7414–7417.

43. K.A. Bulashevich, I.Y. Evstratov, V.F. Mymrin, and S.Y. Karpov, *Phys. Stat. Sol.*, **(c) 4** (2007) 45–48.

Chapter 5

On-Film Formation of Nanowires: A Route to Defect-Free Nanowire Growth and Device Fabrication

Jin-Seo Noh, Wooyoung Shim, Jinhee Ham, and Wooyoung Lee*
Department of Materials Science and Engineering, Yonsei University, 262 Seongsanno, Seodaemun-gu, Seoul 120-749, South Korea
*wooyoung@yonsei.ac.kr

5.1 Introduction

The formation of microscale metal whiskers on the surface of electrical components is a significant technological concern that can result in short circuits or electrical discontinuities. The multilevel metal electrodes in large-scale integration may experience a significant amount of mechanical fatigue owing to the difference in the thermal expansion coefficients of the metals. With repeated device operation, thermally induced mechanical stress can develop to a level that causes substantial atomic diffusion to the extent where it leads to whisker formation. Whiskers are considered a potential threat for device reliability. These whiskers are unintended, are of uncontrollable size, and are limited with respect to the type and

Vistas in Nanofabrication
Edited by Faiz Rahman

ISBN 978-981-4364-56-0 (Hardcover), 978-981-4364-57-7 (eBook)
www.panstanford.com

number of materials that can be grown within microscopic field of view. The whiskers are, therefore, still at present the subject of being remedial studies. From the point of view of materials scientists and engineers, whiskers are ideal for studying material properties, owing to their near perfect single crystal structure afforded by spontaneous thermodynamic growth. Spontaneous stress-induced tin (Sn) whiskers, for example, are single crystals without lattice defects. This was shown by investigation of their elastic and plastic properties [1]. If whiskers with a nanoscale diameter are grown on a macroscopic length scale, it provides one with a platform to fully explore novel fundamental properties which can benefit from the high crystal quality offered by spontaneous growth. In this regard, a recently developed methodology — on-film formation of nanowires (OFF-ON) — has emerged as a promising tool for growing defect-free functional nanowires. This chapter discusses the OFF-ON methodology and its underlying mechanism, focusing primarily on the growth of bismuth (Bi) and bismuth telluride (Bi_2Te_3) nanowires.

5.2 On-Film Formation of Nanowires

The OFF-ON is a stress-induced nanowire growth method that relies on atomic migration under externally applied stress [2]. Compressive stress induced by a thermal expansion mismatch between a film and the substrate is the driving force for the OFF-ON. The OFF-ON requires that the thermal expansion coefficient of a film (α_F) must be larger than that of the substrate (α_S), $\alpha_F > \alpha_S$. In this simple picture, compressive stress is generated in the film by thermal annealing, and the film releases the compressive stress via nanowire growth. This advantage of the OFF-ON bypasses the need for templates, catalysts, and precursors that can be a source of point defects that degrade material properties, thereby avoiding both internal and surface contamination.

As illustrated in Fig 5.1, nanowire growth by the OFF-ON method is simple; it consists of thin-film deposition, annealing, and cooling. In the first step, a thin film, on which nanowires will be grown, is deposited onto a substrate by sputtering. In the next step, the film on the substrate is annealed at an elevated temperature. The annealing temperature is usually set high enough to generate substantial thermal expansion in both the film and the substrate and to facilitate massive atomic diffusion in the film. According to the basic requirements for the OFF-ON method, the thermal expansion of the film is greater than the thermal expansion of the substrate, generating compressive stress in the film. This compressive stress is relieved by nanowire growth at localized positions, usually along grain boundaries. The growth of nanowires is enabled by massive atomic flow at the high annealing temperature. Due to this thermally activated spontaneous mechanism to relax the stress stored in the film, nanowires grown by the OFF-ON method are in general high quality and single-crystalline.

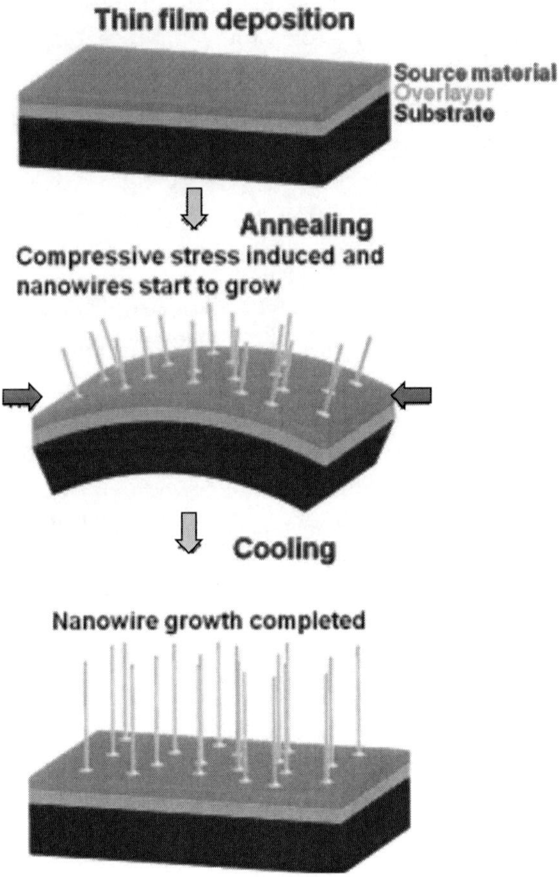

Figure 5.1 Schematic of nanowire growth by the OFF-ON method [2].

5.3 Applicability of the OFF-ON Method to Diverse Nanowires

5.3.1 Bi Nanowires

Bismuth (Bi) is a group V semimetallic element with a small overlap of conduction and valence bands (~38 meV) [3]. This material shows unusual transport properties due to its highly anisotropic Fermi surface, low-carrier concentrations, small-carrier effective mass, and long-carrier mean free path, which have made it well-studied for many applications. Bi nanowires have been intensively researched, particularly in fundamental condensed-matter physics and thermoelectrics. The success of ongoing research on Bi mostly relies on growing high-quality nanowires with sufficient yield.

Bi has several advantages with regard to ease of growth by the OFF-ON method: a low melting point (271.3°C), a high thermal expansion coefficient (13.4×10^{-6}/°C), and a low activation energy for self-diffusion (11.3 kJ/mol). It has been demonstrated that high quality Bi nanowires can be grown by the OFF-ON method. A piece of Si with a SiO_2 overlayer is generally used as the substrate. Its thermal expansion coefficient is (0.5×10^{-6}/°C for SiO_2)/(2.4×10^{-6}/°C for Si), which is more than 20 times smaller than that of Bi. A Bi film is sputter-deposited onto the substrate with a high deposition rate of ~30 Å/s. Depending on the desired nanowire diameter, the film thickness that later determines the diameter of nanowires can be deliberately varied by simply controlling the deposition time, and is usually chosen to be between 40 and 100 nm. The Bi film on the substrate is next transferred into a vacuum furnace and annealed for 5 to 10 hours at ~250°C. During the annealing process, the Bi film expands more than the substrate, generating a substantial compressive stress in the film. To relieve the stress, Bi atoms diffuse through the grain boundaries, and Bi nanowires start to grow at positions where the diffusing atoms penetrate the surface Bi oxide. Figure 5.2 shows bush-like Bi nanowires and a high-resolution view of a nanowire grown by the OFF-ON method. The nanowires are found to be several micrometers in length and several tens of nanometers in diameter (Fig. 5.2a) and are of high quality single-crystalline form (Fig. 5.2c,d).

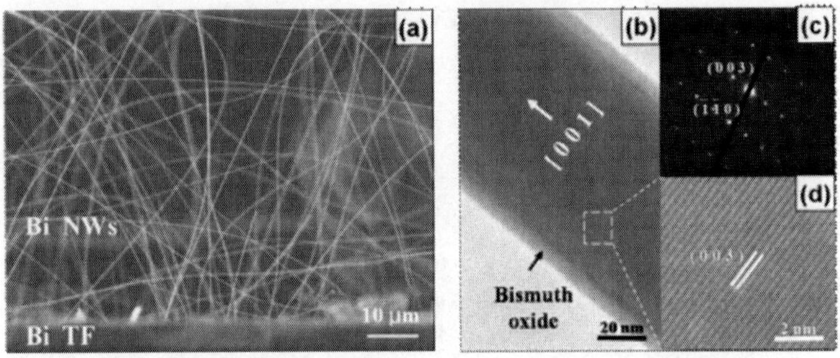

Figure 5.2 (a) A scanning electron microscopy (SEM) image of a bunch of Bi nanowires grown by the OFF-ON method. (b) A low-magnification transmission electron microscopy (TEM) image of a single Bi nanowire. (c) A selected-area electron diffraction (SAED) pattern along the zone axis indicating that the nanowire is highly single-crystalline and its growth direction is [001]. (d) A high-resolution TEM image showing well-aligned atomic planes without defects [2].

To further clarify the growth mechanism, a Bi nanowire of 150 nm in diameter grown on a Bi thin film was selected for cross-sectional analysis [4]. The cross-section dark-field TEM image (low magnification, Fig. 5.3a) shows that the Bi thin film is polycrystalline, consisting of many columnar grains, while the Bi nanowire is a single crystal. Corresponding electron diffraction patterns (inset of Fig. 5.3a) show that the growth direction of the Bi nanowire is [110]. Figure 5.3b shows that the SiO_2 layer beneath the wire is nonplanar, but the thickness of the Bi film in this region is unchanged. An indication of this is that the local stress in this area is sufficient to induce diffusion of SiO_2 away from this region. Grain boundary sliding is also observed, which reduces the stress that drives wire growth. Finally, it is worth noting that Fig. 5.3a shows hillocks as well. According to previous studies [5], the motion of grain boundaries is largely responsible for the growth of whiskers and hillocks. The high mobility of grain boundaries during migration results in broad hillocks on a surface, whereas low mobility or pinned boundaries promotes whisker formation by maintaining a laterally small grain where atom flux accumulates. A grain with a mobile grain boundary can be broadened laterally by grain boundary migration as it is pushed upward, while the other grain broadening can be hindered with pinned grain boundaries that promote nanowire formation.

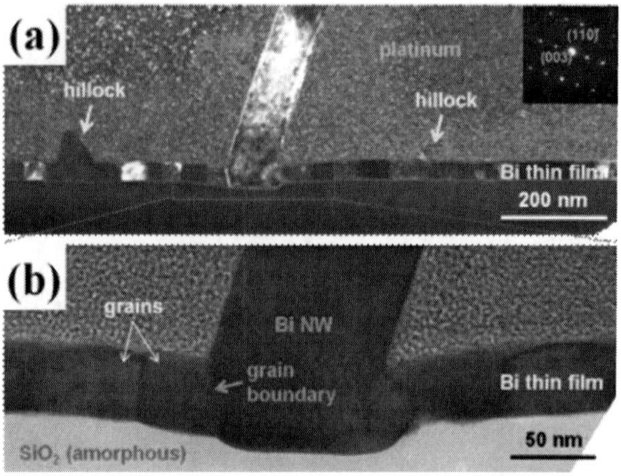

Figure 5.3 TEM images and diffraction pattern of the cross section of a Bi nanowire — Bi thin film. (a) TEM image of the cross section of the Bi nanowire grown on a Bi thin film. (b) Magnified image of the Bi nanowire grown at a grain of the Bi thin film [4].

5.3.2 Bismuth Telluride Nanowires

Bismuth telluride (Bi_2Te_3) is the most widely used thermoelectric material whose thermoelectric figure-of-merit, *ZT*, is approximately 1 at room temperature. It is also recognized as an important topological insulator, which shows nearly perfect electrical conduction on the surface in contrast to insulator-like behavior in the bulk [6]. Similar to Bi, low dimensional Bi_2Te_3 exhibits greatly enhanced thermoelectric performance. Venkatasubramanian *et al.*, for example, demonstrated $ZT \sim 2.3$ for Bi_2Te_3/Sb_2Te_3 superlattices [7]. Low-dimensional Bi_2Te_3 structures are equally important for the study of topological insulators because surface states in low-dimensional structures become significant as the surface-to-volume ratio increases.

The OFF-ON method can be used for the growth of high quality Bi_2Te_3 nanowires. Two different approaches are used: a one-step process and a two-step process. The one-step process refers to the use of the same experimental protocol as for the Bi nanowire growth, as represented in Fig. 5.4a [8]. First, a BiTe thin film is deposited onto a Si substrate with a SiO_2 overlayer, normally by sputtering. Either sputtering from a BiTe compound target or co-sputtering from separate Bi and Te elemental targets can be used for this purpose. The BiTe film thickness is controlled in the range of 40 to 100 nm. In the next step, the BiTe film on the substrate undergoes vacuum-annealing at a high temperature. The annealing temperature should be set higher than that for Bi nanowire growth because of the higher melting point of Bi_2Te_3 (585°C). The annealing temperatures should be between 300°C and 400°C. The BiTe film expands more than the substrate does, as indicated by its thermal expansion coefficient ($\sim 13.4 \times 10^{-6}$/°C) which is larger than the thermal expansion coefficient of the substrate ((0.5×10^{-6}/°C)/(2.4×10^{-6}/°C)). The compressive stress induced in the BiTe film is the driving force for Bi_2Te_3 nanowire growth, as it is for Bi nanowire growth. It has been demonstrated that Bi_2Te_3 nanowires grown by this method are highly single-crystalline and uniform in diameter without tapering. The growth mechanism and characterization of Bi_2Te_3 nanowires grown by the OFF-ON method is described in Ref. [8].

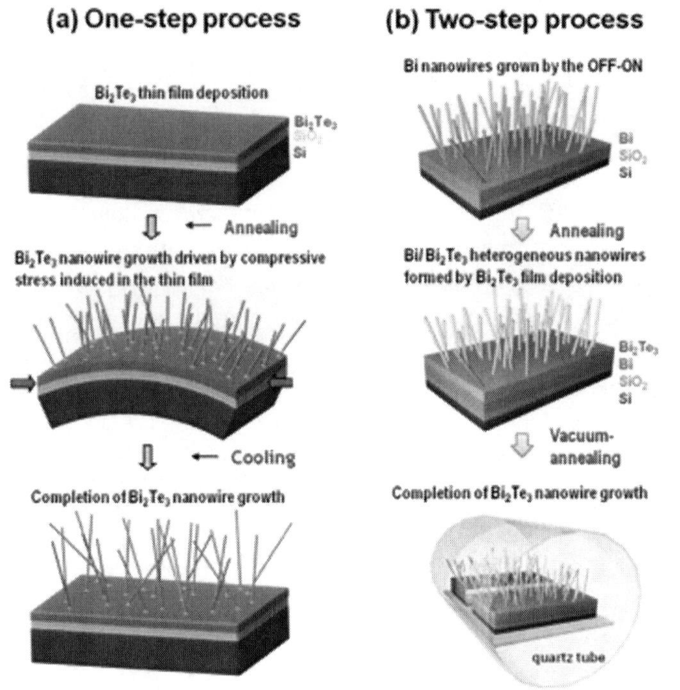

Figure 5.4 Schematic representation of the Bi_2Te_3 nanowire growth process: (a) one-step process, (b) two-step process [8, 9].

Another approach for growing Bi_2Te_3 nanowires is the use of core-shell architecture that is then subject to the interdiffusion of Bi and Te atoms under annealing process. In this process, schematically illustrated in Fig. 5.4b, Bi nanowires are grown first using the OFF-ON method, and then a Bi_2Te_3 thin film is sputter-deposited onto the Bi nanowires. The Bi nanowires are then coated with a Bi_2Te_3 thin film. These heterogeneous Bi/Bi_2Te_3 core-shell nanowires are transformed into homogeneous Bi_2Te_3 nanowires by massive interdiffusion of Bi and Te atoms during another annealing step between 300°C and 400°C. The excess Bi atoms are vaporized from the nanowire surface, making the nanowires thermodynamically stable with regard to chemical composition. The Bi_2Te_3 nanowires grown by this two-step process appear to be uniform in diameter and are single-crystalline.

5.4 Control over the Size and Density of Nanowires

5.4.1 Nanowire Diameter

The utility of the OFF-ON method as a tool for growing nanowires relies on achieving control over the nanowire diameter, which determines the physical properties of the nanowires. In previous studies of whisker formation, the goal was to prevent them or remove them, and thus controlling the diameters of whiskers was not the topic of study. In this section, factors affecting the size and density of nanowires grown by the OFF-ON method are discussed.

The key factor determining the average diameter of the nanowires is the average grain size [2], as shown in Fig. 5.5. The Bi nanowire diameter decreases from 450 nm to 98 nm as the thickness of the Bi film decreases from 830 nm to 55 nm. The average grain size of a Bi film becomes smaller as the Bi film thickness decreases. These correlations between nanowire diameter, film thickness, and grain size appear to be valid over the Bi film thickness range of 40 nm to 3.3 μm. This indicates that the decrease in grain size of a Bi film results in a decrease of the diameter of the Bi nanowires. To further reduce the nanowire diameter, even smaller grains are required. In fact, the smallest diameter of Bi nanowires that can be grown from a Bi film deposited at room temperature is 50 nm. In the context of reducing grain size in a Bi film, a Bi film was deposited at low temperatures where grain size is expected to be small. Bi nanowires with diameters as small as 20 nm were grown by depositing a Bi film at 77 K.

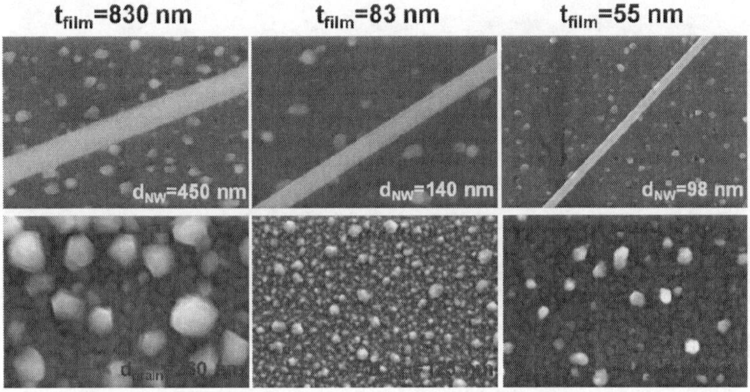

Figure 5.5 SEM images of Bi nanowires (upper row) grown on Bi films (bottom row) deposited with different thicknesses at room temperature [2].

5.4.2 Nanowire Length

Annealing temperature and time are the key factors to control the length of a nanowire grown by the OFF-ON method. This is clearly demonstrated for Bi nanowires (Fig. 5.6) that is imaged in real time by SEM. It is noted that the length of the Bi nanowire gradually increases with increasing annealing temperature and time. This is reasonable because the OFF-ON is a nanowire growth method based on diffusional mass flow, which is more highly activated at higher temperatures, and the total amount of mass flow is nearly proportional to the duration of a sample at a given temperature. As can be found in Fig. 5.6, a very long annealing time may be needed at low temperatures to grow nanowires to a desired length. Therefore, the moderately high annealing temperatures close to the melting point of Bi are desirable to shorten the annealing time and increase the crystallinity of resultant nanowires.

Figure 5.6 SEM images of Bi nanowires grown at different annealing temperatures for varying annealing times [4].

5.4.3 Nanowire Density: Type of Substrate

The OFF-ON method is a powerful tool for growing a variety of high quality nanowires, but the resultant nanowire density by this method is generally low. One of the challenges facing the OFF-ON at present, therefore, is to increase nanowire density in a controlled and practical fashion. For this reason, it is important to consider the factors necessary to achieve high nanowire density. Because thermal expansion difference between a film and the substrate provides a driving force for nanowire growth, the selection of suitable substrates provides an effective strategy for optimizing the condition towards this goal. As discussed in Section 5.2, the difference in thermal expansion coefficients of a film and the substrate needs to be large, which in turn induces a large magnitude of compressive stress in the film. For comparison, Bi nanowire densities grown using Si with a SiO_2 overlayer (α = 0.5 × 10–6/°C) and pure Si (α = 2.4 × 10–6/°C) substrates are presented in Fig. 5.7. The nanowire density from SiO_2/Si substrate is found to be more than an order of magnitude larger than that from Si substrate when all other growth conditions are the same. An indication of this result is that the large variety of different substrates with lower thermal expansion coefficients can be used to increase nanowire density.

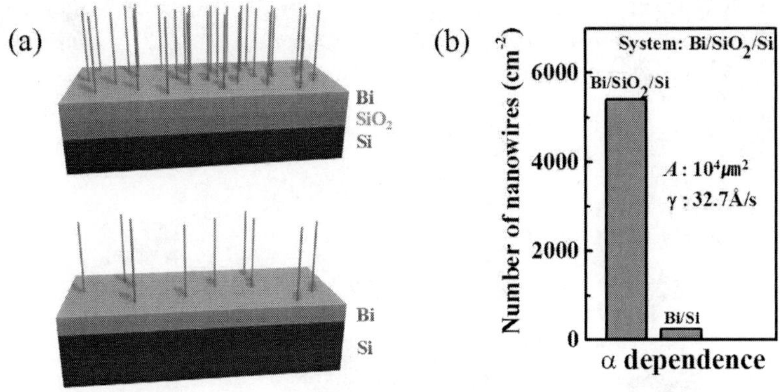

Figure 5.7 (a) Schematic drawings and (b) measured densities of Bi nanowires grown using SiO_2/Si and pure Si substrates with setting all other conditions as the same for the two cases [10].

5.4.4 Nanowire Density: Film Deposition Rate

Another significant factor contributing to nanowire density is film deposition rate. It is because the morphology and grain structure of a film strongly depend on the film deposition rate. When a film is deposited by sputtering at a fixed temperature, the impinging flux of constituent atoms is higher at a high deposition rate. This indicates that the average interval between encounters of adatoms is shorter at the high deposition rate, forming local excesses of adatoms that are thermodynamically unstable. They are inclined to migrate upon thermal annealing. The high deposition rate is also expected to cause a finer grain structure as adatoms are hardly able to move to their stable positions before the following flux of adatoms arrives at the surface of growing film. The finer grain structure can stimulate nanowire growth at many dispersed points, through efficient delivery of raw material along enlarged grain boundaries. These effects of film deposition rate are demonstrated for Bi nanowires grown on SiO_2/Si substrate in Fig. 5.8. From the comparison of Fig. 5.8(a) and (b), it is found that the surface of a Bi film deposited at a high deposition rate (32.7 Å/s) is rougher than that of a slowly grown Bi film with many protrusions on it. The surface protrusions may play a role as seeds for nanowire growth since compressive stress can be more easily relieved at these protruded points where Bi oxide is less likely to form on their surfaces or is easily broken. As a consequence, the nanowire density from a Bi film deposited at 32.7 Å/s appears to be more than 700 times higher than from its counterpart deposited at 2.7 Å/s.

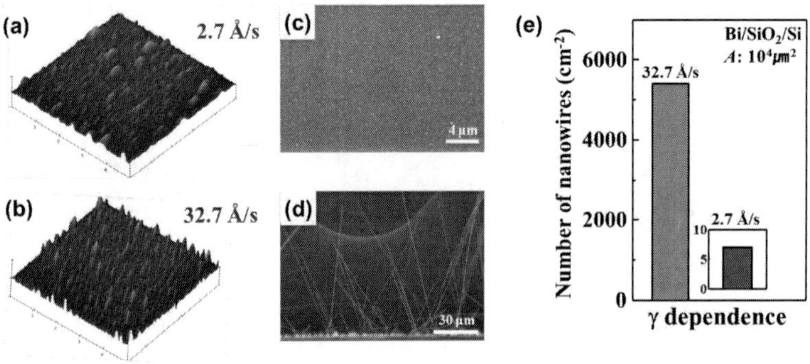

Figure 5.8 Atomic force microscopy (AFM) images of 50 nm-thick Bi films deposited with the rate of (a) 2.7 Å/s and (b) 32.7 Å/s on SiO_2/Si substrate. (c, d) are SEM images of Bi nanowires grown from the respective films, and (e) shows the difference in nanowire densities for the two cases [10].

5.4.5 Nanowire Density: Film Area and Substrate Patterns

Because the OFF-ON method utilizes a thermally induced compressive stress, the nanowire density can be affected by related factors influencing the magnitude of stress caused in a film. The size of a film is a typical example. The nanowire density is expected to decrease as the film area shrinks, as more compressive stress is likely to be released at the edge of a film. This film size dependence of nanowire density can be confirmed by nanowire growth on patterned films with different area sizes. Figure 5.9 shows that Bi nanowire density gradually decreases with decreasing the pattern area from $(10^4\ \mu m)^2$ to $(1\ \mu m)^2$. The additional study, in which Bi films were deposited into pre-patterned SiO_2 sidewalls, has resulted in significantly enhanced nanowire density by a factor of 7 as compared to non-patterned large area Bi films [11]. The pre-patterned SiO_2 surfaces that are designed to suppress the relaxation of the film stress serve to increase the nanowire density.

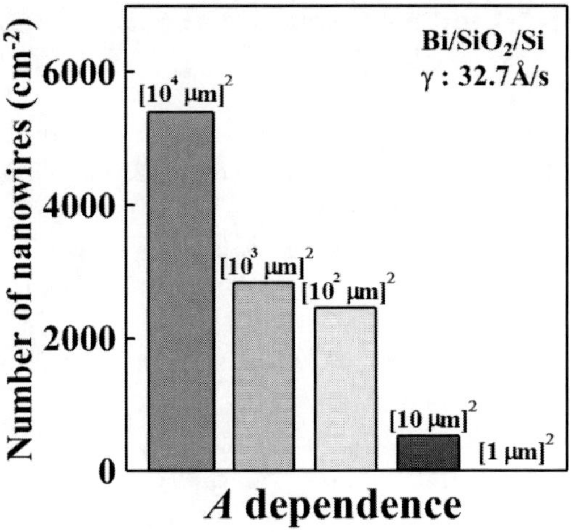

Figure 5.9 Densities of Bi nanowires grown on Bi patterns with different areas [10].

5.5 Self-Assembled Lateral Interconnects

5.5.1 Needs for Self-Assembled Lateral Interconnection

The primary challenges facing the electronics community at present is to increase the integration density of nanoscale structures by using top-down approaches. However, the conventional top-down approaches are expected to face their limits beyond a certain feature size. To address this potential issue, various bottom-up approaches have been developed and showed the potential as a tool, but these still suffer from several issues. Positioning of individual nanostructures with precise control is the main requirement for their success. Lateral interconnects are a common form of electronic architecture that essentially needs control over position and dimension. Various self-assembly techniques have been proposed, including SiO_2 patterning and cracking [12], field-emission induced growth by scanning tunneling microscopy [13], and bridging from each patterned catalyst island [14]. However, the techniques are complicated and costly. Therefore, there is a growing demand for establishing a self-assembly strategy that is useful for easily forming high-quality lateral interconnects between nanostructures.

5.5.2 Self-Assembled Lateral Interconnection Using the OFF-ON Method

Compressive stress is accumulated in a film in the conventional OFF-ON method. To relieve this horizontal compressive stress, nanowires grow perpendicular to the film plane. The situation is different if a film is capped with another film having a low thermal expansion coefficient. As schematically shown for lateral Bi nanowires in Fig. 5.10, the film is now subject to the second compressive stress that acts perpendicular to the plane. This perpendicular stress is generated by suppression of vertical expansion of the film situated in the sandwich structure. In addition, the capping layer blocks diffusion flow for nanowire growth occurring normal to the plane. Under these circumstances, the perpendicularly accumulated stress is relieved through lateral mass flow, leading to lateral nanowire growth. Since the film thickness is usually far thinner than the size of a pattern, the magnitude of perpendicular stress that should be released through unit area of film side is larger than that of horizontal stress per unit area of film top. Besides, lateral mass supply may be more facilitated, once laterally grown nanowires are rare. These presumably lead to larger nanowire diameters than the ones grown vertically.

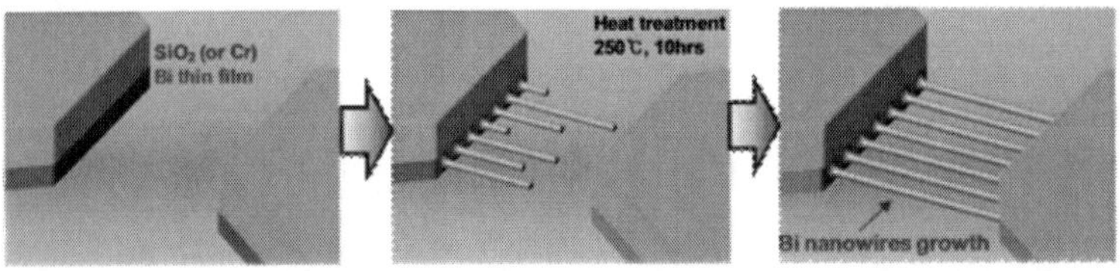

Figure 5.10 Schematic pictures of lateral Bi nanowire growth process. Step 1: pre-patterning a Cr (or SiO_2)/Bi bilayer structure on a thermally oxidized Si substrate. Step 2: initiation of lateral Bi nanowire growth during annealing at 250°C for 10 hours. Step 3: completion of lateral Bi nanowire growth [15].

5.5.3 Lateral Bi Interconnection Between two Cr/Bi Electrodes

Following the basic mechanism explained above, lateral Bi nanowires are indeed demonstrated to grow and bridge between pre-patterned Bi films that are capped with a chromium (Cr) film. The proof-of-concept experimental result is shown in Fig. 5.11a. A lateral Bi nanowire bridges two Cr/Bi patterns across a 1 μm-deep, 4 μm-wide trench. Here, the Cr capping layer plays a role as a source of the second compressive stress and Bi diffusion barrier in the perpendicular direction. The diameter of the lateral nanowire is 192 nm, which is more than two-fold thicker than the conventional vertical Bi nanowires grown with the same growth conditions. Similar to vertical nanowire growth, the diameter of lateral nanowires can be controlled within a certain extent by adjusting the film thickness.

Figure 5.11b exhibits an SEM image of advanced Cr/Bi electrode patterns that are designed to be used for self-assembled lateral nanowire growth. The electrode ends are sharply tapered at two sides where lateral nanowires are likely to be grown preferentially between the sharp tips, as compressive stress developing during annealing will concentrate at the tips. Indeed, lateral Bi nanowires spontaneously grow between adjacent tips to connect neighboring Cr/Bi electrodes, as shown in Fig. 5.11c.

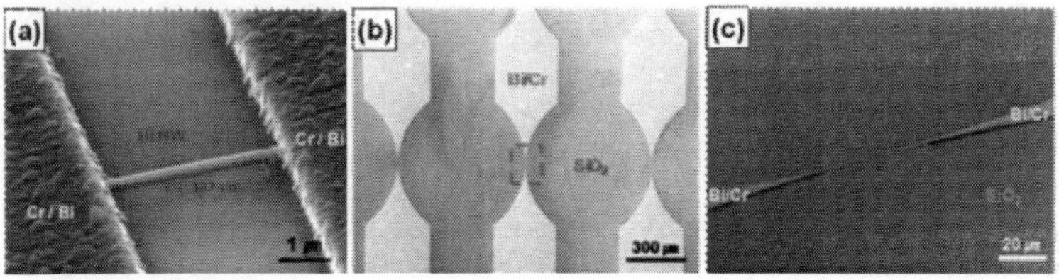

Figure 5.11 SEM images of (a) a lateral Bi nanowire grown across two parallel Cr/Bi electrodes 4 μm distant from each other, (b) Bi/Cr electrode patterns with sharp tips, and (c) a lateral Bi nanowire grown between the neighboring sharp tips. Lateral Bi nanowires were grown by annealing the patterns at 250°C for 10 hours, using SiO_2/Si substrate [15].

5.5.4 Electrical Properties of Lateral Bi Interconnects

Making Ohmic contacts between Bi and metal electrodes has been a problem for investigating novel transport properties in Bi, largely due to an oxide layer on Bi surfaces. The Bi interconnects grown by the OFF-ON method, combined with self-assembly strategy, appear to be highly Ohmic based on the observation of linear current (I)–voltage (V) characteristics, as shown in Fig. 5.12a. When considering that the top-down approaches for patterning electrodes on an individual nanowire are difficult and hardly result in good Ohmic contacts, this result simply manifests the efficiency of self-assembled interconnections between pre-patterned electrodes.

Figure 5.12b shows a magnetic field (H)-dependent transverse magnetoresistance (MR) for a 600 nm thick lateral Bi nanowire at 300 K. Here, MR is defined as $\Delta R(H) / R(0) = [R(H) - R(0)] / R(0)$, where $R(0)$ and $R(H)$ are the resistance at zero field and at a given magnetic field, respectively. A large transverse MR of 123% is observed for this self-assembled nanowire. The large MR is caused by long carrier mean free path in a high quality Bi nanostructure and validates the high crystallinity of the lateral Bi nanowire.

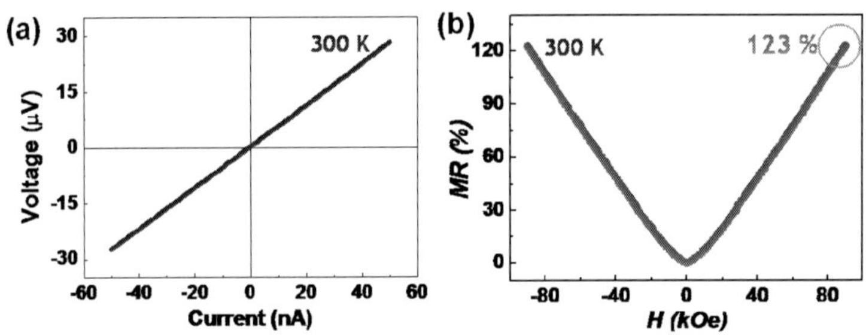

Figure 5.12 (a) I–V characteristic for a Bi interconnect at room temperature. (b) Transverse MR vs. magnetic field for a 600 nm-thick lateral Bi nanowire at room temperature [15].

5.6 Summary

The OFF-ON is the novel method that retains many of the advantages of whisker formation – high crystallinity, fast growth rate with uniform diameter, and a facile route to produce nanowires – while overcoming the issue of precise control over diameter that previously has not been achieved by conventional whisker formation. The method combines the concepts and advantages of simple thin film deposition with whisker formation that produces highly crystalline nanowires. Unlike other whisker formation methods, the OFF-ON offers significant advantages of providing the tunability of nanowire diameter, e.g., Bi as small as a few tens of nanometers and enabling the synthesis of thermodynamically stable compound semiconductors, e.g., Bi_2Te_3. Furthermore, the OFF-ON-based self-assembled interconnection is demonstrated, which can be recognized as a practical advancement in bottom-up approaches, as it addresses the issue of nanowire position controllability over the length scale that is required for high density integration.

Although advances have been rapid and span many concepts, several aspects of the OFF-ON continue to challenge researchers. The range of materials readily available for nanowire growth by the OFF-ON must be expanded before the full potential of this technology is realized. Additionally, the use of this technique as a tool for rapid and massive fabrication for functional nanowires and devices that could transform the manufacturing way of integrated circuits requires a route to site-specific functional nanowire synthesis. Regardless of these challenges, the OFF-ON method has become an affordable and reliable technology available to researchers in nanoscience and nanotechnology community.

References

1. Herring, C., and Galt, J.K. (1952). Elastic and plastic properties of very small metal specimens, *Phys. Rev.,* **85**, pp. 1060–1061.
2. Shim, W., Ham, J., Lee, K.I., Jeung, W.Y., Johnson, M., and Lee, W. (2009). On-film formation of Bi nanowires with extraordinary electron mobility, *Nano Lett.*, **9**, pp. 18–22.
3. Zhang, Z., Sun, X., Dresselhaus, M.S., Ying, J.Y., and Heremans, J. (2000). Electronic transport properties of single-crystal bismuth nanowire arrays, *Phys. Rev. B,* **61**, pp. 4850–4861.
4. Ham, J., Shim, W., Kim, D., Oh, K., Voorhees, P.W., and Lee, W. (2011). Watching bismuth nanowires grow, *Appl. Phys. Lett.,* **98**, pp. 043102.
5. Boettinger, W., Johnson, C. Bendersky, L., Moon, K., Williams, M., and Stafford, G. (2005). Whisker and hillock formation on Sn, Sn–Cu and Sn–Pb electrodeposits, *Acta Mater.,* **53**, pp. 5033–5050.
6. Moore, J. (2009). Topological insulators: The next generation, *Nat. Phys.*, **5**, pp. 378–380.
7. Venkatasubramanian, R., Siivora, E., Colpitts, T., and O'Quinn, B. (2001). Thin-film thermoelectric devices with high room-temperature figures of merit, *Nature*, **413**, pp. 597–602
8. Ham, J., Shim, W., Kim, D.H., Lee, S., Roh, J., Sohn, S.W., Oh, K.H., Voorhees, P.W., and Lee, W. (2009). Direct growth of compound semiconductor nanowires by on-film formation of nanowires: Bismuth telluride, *Nano Lett.*, **9**, pp. 2867–2872.
9. Kang, J., Noh, J.S., and Lee, W. (2011). Simple two-step fabrication method of Bi_2Te_3 nanowires, *Nanoscale Res. Lett.*, **6**, pp. 277.

10. Shim, W., Ham, J., Noh, J.S., and Lee, W. (2011). Structure-dependent growth control in nanowire synthesis via on-film formation of nanowires, *Nanoscale Res. Lett.*, **6**, pp. 196.

11. Kim, H., Noh, J.S., Ham, J., and Lee, W. (2011). Promoted growth of Bi single-crystalline nanowires by sidewall-induced compressive stress in on-film formation of nanowires, *J. Nanosci. Nanotechnol.*, 11, pp. 2047–2051.

12. Alaca, B.E., Sehitoglu, H., and Saif, T. (2004). Guided self-assembly of metallic nanowires and channels, *Appl. Phys. Lett.*, **84**, pp. 4669–4671.

13. Thong, J.T.L., Oon, C.H., Yeadon, M., and Zhang, W.D. (2002). Field-emission induced growth of nanowires, *Appl. Phys. Lett.*, **81**, pp. 4823–4825.

14. Franklin, N.R., Wang, Q., Tombler, T.W., Javey, A., Shim, M., and Dai, H. (2002). Integration of suspended carbon nanotube arrays into electronic devices and electromechanical systems, *Appl. Phys. Lett.*, **81**, pp. 913–915.

15. Ham, J., Kang, J., Noh, J.S., and Lee, W. (2010). Self-assembled Bi interconnections produced by on-film formation of nanowires for *in situ* device fabrication, *Nanotechnology*, **21**, pp. 1653021–1653025.

Chapter 6

Nanotrenches: An Optical Lithography Process for High-Aspect-Ratio sub-100 nm Gaps

Jean-Francois Dayen,* Vina Faramarzi, and Bernard Doudin

IPCMS-Department of Magnetic Objects on the Nanoscale,
Institut de Physique et Chimie des Matériaux de Strasbourg (UMR 7504 CNRS),
Université de Strasbourg, 23 rue du Loess, BP 43,
Strasbourg, F-67034 Strasbourg Cedex 2, France.
*jean-francois.dayen@ipcms.u-strasbgunistra.fr

We present a simple and versatile patterning procedure for the reliable and reproducible fabrication of high-aspect-ratio (10^4) electrical interconnects that have separation distances down to 20 nm and lengths of several hundreds of microns. The process uses standard optical lithography techniques and allows parallel processing of many junctions, making it easily scalable and industrially relevant.

To illustrate the advantage of the nanotrench architecture for nano and molecular electronics applications, two different kinds of working devices are presented, spanning the ranges from high-density current systems, up to the mA range, to highly resistive nanoparticles network with sup pA current, and from micro down to nano interconnects systems. Trapping of ferromagnetic

Vistas in Nanofabrication
Edited by Faiz Rahman

ISBN 978-981-4364-56-0 (Hardcover), 978-981-4364-57-7 (eBook)
www.panstanford.com

intermediate scale particles for multiscale interconnects purposes is also presented.

6.1 Introduction

New breakthroughs envisioned for molecular and nano electronics in the area of information storage and processing [1,2] have motivated a significant boost to both fundamental and applied research in the area. Electronic transport through nanoscale down to molecular entities can potentially take advantage of exciting multi-functional properties, with the expectation of dominant quantum mechanical effects and original chemical tuning opportunities. As the field is becoming more mature, it is becoming increasingly clear that the realization of robust single molecule or single nanoparticle devices exhibiting reliable and reproducible behavior is a very challenging experimental and technological task [3–7]. Among the main problems to tackle, which severely limit the possibility of achieving reliable single entity-based devices for scaling up industrial production, are, for instance, their high sensitivity to chemical and electromagnetic environment [8,9], the critical role played by defects or uncontrolled doping in the electronic transport properties [10,11] or the control at nanometer scale precision of the positioning of a single molecular-sized entity in an electronic circuitry. The development of single molecule-based electronic devices for industrial purposes seems, therefore, premature at this time.

A possible route to reconcile the molecular scale (below 1 nm) and the top-down nanofabrication route (larger than 10 nm) is to use a multi-scale approach, where a cluster (metallic, semimetallic, or semiconductor) of typically 10 nm size [12] or a conducting microsphere, of typical 10^2–10^3 nm size, is used as an intermediate bridge. One can also take advantage of bottom-up fabrication methods for incorporating macroscopic amounts of identical molecules into devices or to assemble well-organized monolayers of nanoparticles [13–15]. Key advantages are simplicity, ability to cover large areas, and cheap manufacturing costs. Large numbers of interconnected molecules also ensure device robustness through ensemble averaging [16,17]. Nanoscale entities such as functionally ordered arrays of nanoparticules also represent an alternative opportunity for designing new devices such as gas sensor, electro-optic devices, or magneto-electronic nanodevices [18–20]. The transport and magneto-transport mechanisms taking place are often very sensitive to electrode/nanoparticle molecular interface, which can create artifacts or difficulties in unambiguous interpretation of the data. Devices based on multiple connected nanoparticles should be less sensitive to such artifacts and improve the reproducibility of the measurements from one sample to the other. Hence, there is a need for a versatile configuration of device, compatible with bottom-up fabrication methods and made by standard optical lithography techniques to achieve low manufacturing costs.

Planar electrode geometry, directly inspired by top-down nanofabrication techniques, provides several key advantages. First, the integration of a three-terminal configuration device capable of gating the flow of carriers can be easily implemented. Furthermore, planar substrates are also ideally compatible with bottom-up fabrication methods involving nanoparticle layers such as Langmuir-related techniques. Finally, the open direct access to the middle molecules or nanoparticle layers optimizes chemical and optical control or excitation. This is particularly attractive if we want to gain access to the intrinsic properties of the molecules for the purpose of realizing multi-functional devices. Lateral geometry also provides unique advantages versus the more ubiquitous vertical geometry [21–23], where metal electrodes sandwich an organic film. Such geometry suffers from the key problem of possible formation of leaking metallic paths during the deposition of the top electrode [24,25], which can only be avoided by adding an intermediate highly conductive polymer film [16].

We present a recently developed process [26] to design wide electrodes (potentially up to mm scale), separated by a short distance (down to 20 nm) corresponding to typical thicknesses in organic vertical structures. There is an extensive literature on nanofabrication techniques for realizing closely spaced electrodes, and it is clear that such dimensions should be easily achievable by e-beam lithography tools. One should, however, emphasize that the critical issue of obtaining high-aspect-ratio (HAR) gaps below 50 nm is an extremely challenging task, necessitating state-of-the-art electron beam equipment, which might suffer from reliability issues when extremely high aspect ratios are desired. More importantly, the use of a sacrificial layer becomes necessary if aspect ratio exceeding 10–100 are needed. Layers made of aluminum oxide [27], chromium oxide [28] or molecular monolayers [29] are reported in the literature. All these techniques suffer from the requirement to have a robust etching process, which needs to be critically selective and critically efficient.

Realizing HAR gap devices becomes of key importance when materials with intrinsically large resistivity are investigated. In the field of nanoelectronics, materials with high sheet resistance values are often encountered. Hopping transport makes low-temperature measurements critical and becomes a practical limitation to studies. Conversely, when low resistance samples are investigated, series resistance of the leads, and their intrinsic resistivity, might limit the measurement quality and reliability. There is, therefore, a need for realizing planar electrode interconnects with large aspect ratios, exhibiting good interface resistance, and compatibility with low and high impedance measurement systems.

One industrially attractive approach that does not rely on electron beam lithography, maximizes the reliability for HAR trenches, and minimizes contamination is the so-called shadow edge evaporation method [26,30]. This technique has been used previously to make nanogaps of very small widths (around 100 nm) to contact single entities [31,32] or to make a template for nanowire and nanochannel patterning [30,33]. We have developed a process [26] using only standard optical lithography to make HAR gap devices consisting of two planar electrodes separated by a "nanotrench" 20 to 100 nm wide and up to 100 μm in length, i.e., with a ratio up to 1:5000. This ratio can be straightforwardly increased by 1–2 orders of magnitude by using an inter-digitized geometry or by patterning longer electrodes. This is because the process is based on a simple geometrical effect which can be effective for any length. It would be only limited by the optical lithography capability and intrinsic wafer defects. Avoiding the use of more sophisticated lithography techniques makes our fabrication strategy more relevant for industrial up-scaling.

Our goal is to provide compelling evidence of the adequacy of the shadow-edge evaporation approach for making trenches with no detrimental structural defects for applications in electronics. We demonstrate the interest and reliability of this approach by presenting two examples of applications for nanoparticle and microparticle electrical properties, showing that no leakage or shorts are present in our structures, and that the electrical contacts are of good quality and suitable for a wide range of current measurements, going from sub-pA measurements up to tens of mA currents.

6.2 Experimental Details

6.2.1 Fabrication of the Electrodes

Edge-mediated shadow mask lithography allows sample fabrication in only two lithographic steps. Micrometer-scale electrodes (Fig. 6.1a) are made of Ti/Au bilayer (with, for example, typically 3 nm of Ti and 40 nm of Au) deposited using electron beam evaporation followed by a standard lift-off process. The first layer is patterned on silicon substrates covered with thermal oxide by 3.6 seconds of exposure through a positive mask over AZ5214 photoresist followed by 35 seconds flood exposure for image reversal. AZ726 MIF is used as a developer. After metal deposition, the lift-off process is performed in acetone. The second layer is made of lines crossing perpendicularly the first layer electrodes (Fig. 6.1b). Angle deposition (for example 65° for a gap of 80 nm) of metallic Ti/Au bilayer is performed in a chamber with a large distance separating the source and target (Fig. 6.1c).

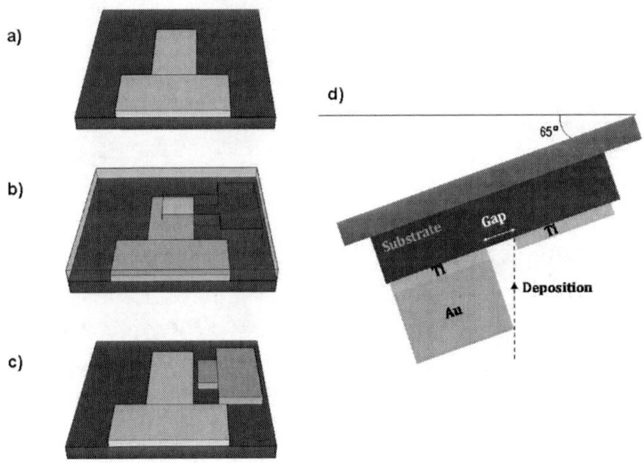

Figure 6.1 Schematics of the edge-mediated shadow mask lithography process. (a) A first pattern is made by standard optical lithographic process, followed by evaporation of a metallic layer and standard lift-off process. (b) The second layer is made by the same method, with patterned lines perpendicular to the first pattern. (c) Metal deposition with the sample tilted at a specific angle is done to form the second metallic electrode and the nanotrench. (d) Schematic of the edge-mediated shadow mask lithography process. The width of the gap is set by the height h of the first electrode and the angle $\theta = 65°$ of the tilted sample.

6.2.2 Nanoparticle Synthesis

Iron oxide nanoparticles are used as model system for potentially highly resistive materials. The iron oxide nanoparticles are obtained by thermal decomposition of an iron stearate precursor in a high boiling point solvent. 1.38 g (2.2×10^{-3} mol) of Fe(stearate)$_2$ (99%, Strem Chemicals) and 1.24 g (4.4×10^{-3} mol) of oleic acid (99%, Alfa-Aesar) are added to 20 mL octyl ether (99%, Sigma-Aldrich, b.p. 287°C).[15] The mixture is heated to 100°C and stirred at this temperature to dissolve the reactants. The solution is then heated to 287°C at a rate of 5°C/min without stirring and refluxed for 120 min under air. The resultant black solution is cooled down to room temperature. The nanocrystals are then washed three times by the addition of ethanol followed by centrifugation (8000 rpm, 10 min). A size-selection procedure is applied by Size Selective Precipitation (SSP): the nanoparticles are suspended in hexane at a concentration of 1 mg/mL and precipitated using the same volume of acetone. After centrifugation, the precipitate is collected and redispersed in THF. The SSP reduces the size polydispersity from 15% to 9%. The suspension is finally dialyzed for three days in THF with a 25 kDa regenerated cellulose membrane (SpectrumLabs) in order to remove any impurities or oleic acid molecules excess, possibly still present after the first purification step. The nanocrystals covered with oleic acid can be easily suspended in various organic solvents (chloroform, toluene, hexane, dichloromethane, etc.).

6.2.3 Langmuir–Blodgett Film Preparation

A small volume (150 μL) of a 5 mg/mL iron oxide nanoparticles suspension in chloroform is spread on the water subphase of the Langmuir trough (KSV 5000, 576 × 150 mm) at room temperature. The area available to nanoparticles is reduced by compressing the barriers (compression rate: 5 mm/min), and the surface pressure-area isotherm is recorded during the film compression using a Wilhelmy plate. Multilayer films are transferred from the water subphase to the substrate supporting the electrodes at a pressure of 30 mN/m by pulling the substrate several times out of the water subphase at a rate of 1 mm/min.

6.2.4 Microsphere Preparation

Sicastar® Monodisperse silica Microspheres, mean diameter ranging from 100 nm to 1.00 μm (Micromod), are diluted and washed in distilled water by centrifuging steps. After standard organic acetone/ethanol cleaning of microscope slides, followed by N_2 drying, the glass surface is made hydrophilic using oxygen plasma. The microspheres in aqueous solution are then spread onto the hydrophilic surface and dispersed uniformly after solvent evaporation. These substrates are covered by e-beam evaporation of 40 nm thick gold films deposited over 3 nm of titanium (as an adhesion layer). The spheres are then detached from the surface by soft sonication in ethanol, washed, collected by centrifuging, dispersed in ethanol, and then spread on the nanotrenches for bridging the electrodes.

6.2.5 Electrical Measurements

Electrical measurements are carried out inside a vacuum-pumped cryostat ($P < 10^{-6}$ mbar) that can reach liquid helium temperature. Measurements of the DC electrical properties are performed using a semiconductor parameter analyzer (Agilent E5270B). Room temperature measurements are performed on a probe station with a four probes configuration (Karl Suss PM8).

6.3 Results and Discussion

6.3.1 Nanotrench

With the approximation of a collimated source, the shadowing results in a gap of size $G = h * \tan(\theta)$, determined by the evaporation angle θ and the height h of the first electrode layer (Fig. 6.1d). It is important to highlight that during the second lithographic step, the nanotrench area is not exposed to reticulated resist. This is of critical importance if molecular electronic studies are envisioned, as possible residue from the reticulated resist after lift-off could interfere with the molecular signature under study or behave as a small current leakage path.

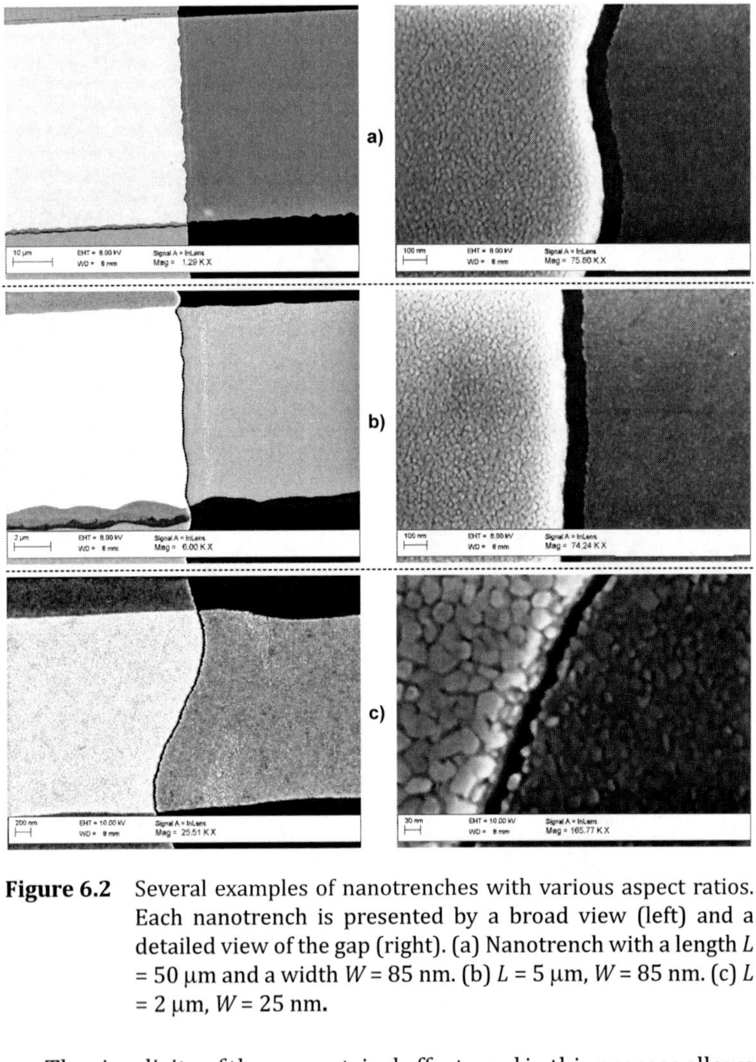

Figure 6.2 Several examples of nanotrenches with various aspect ratios. Each nanotrench is presented by a broad view (left) and a detailed view of the gap (right). (a) Nanotrench with a length L = 50 μm and a width W = 85 nm. (b) L = 5 μm, W = 85 nm. (c) L = 2 μm, W = 25 nm.

The simplicity of the geometrical effect used in this process allows us to realize nanotrenches that are reproducible, uniform all along the micrometric length of the gap, and have no resist contamination inside the gap area. Gaps down to 20 nm on several micrometers long length (Fig. 6.2c) are reproducibly obtained, locally reaching distances of the order of 10 nm. We routinely made several tens of samples with gap sizes below 100 nm and lengths ranging from 2 to 100 μm (see Fig. 6.2, for examples, of HAR gaps of different lengths). These nanotrench devices were obtained successfully on oxidized silicon wafers and on glass or quartz substrates compatible with optoelectronic measurements. Leakage currents below 1 pA for applied bias up to 1 V are systematically found (Fig. 6.3) and considered as benchmark values for high-electrical quality samples. Trenches of gap values down to the grain size of the deposited material, i.e., 10 nm, can be realized. But we observed somewhat poorer reliability, as the majority of the samples with aspect ratios larger than 10^3 with such sub-10 nm gap exhibit leakage currents in the nA range. Such borderline samples are, therefore, not considered but reveal that critical assessment is provided by basic electrical characterization.

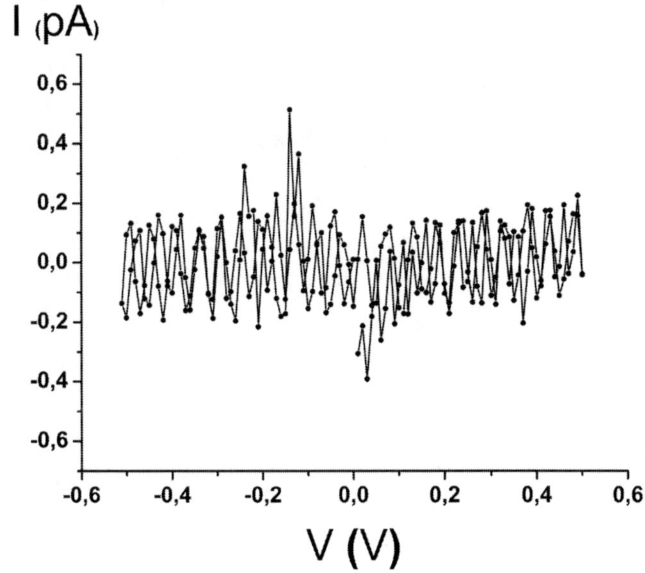

Figure 6.3 Sub-pA leakage current systematically measured on empty nanotrenches.

6.4 Demonstration Devices

6.4.1 Highly Resistive Magnetic Nanoparticles Arrays for Magnetoresistive Devices

In the first example of device fabrication, monolayer and multilayer films of iron oxide nanoparticles are deposited by Langmuir–Blodgett method over the trenches. Films of magnetite-based NPs exhibit very interesting magneto-resistance values when compared to conventional films prepared by physical methods and can potentially lead to the development of devices with enhanced magneto-transport properties [23,34,35]. In this study, initially published in Ref. [26] and extended in Ref. [36], we used magnetite based NPs made of conductive magnetite core and an oxidized shell. These provide intrinsically high resistance, to demonstrate the adequacy of the nanotrench pattern for measurements on intrinsically poorly conducting systems.

The iron oxide nanoparticles have an average size of 11 nm. The nanotrench is oriented along the axis of pulling during the deposition process. SEM images of monolayers and multilayers of iron nanoparticles successfully deposited into the gap are presented in Fig. 6.4a,b, respectively. Once the magnetite film is assembled, oxygen plasma exposure is used to improve the electrical contact between NPs. The resistance R decreases dramatically from the TΩ range corresponding to the empty nanotrench to a value of a few MΩ of magnitude comparable to previous results obtained on samples in vertical geometry [23]. The observed dramatic current increase provides immediate indication of occurrence of conductive path through the sample. The stability and the reproducibility of these observations confirm that the planar geometry is well-suited for compatibility with wet chemistry film deposition processes such as Langmuir Blodgett techniques.

We checked the stability of the electrical contact by cooling the sample down to 50K. Strong non-linear I(V) behavior was systematically observed from 300 K down to 50 K (Fig. 6.4c,d), which may be a manifestation of tunneling through arrays of barrier in the presence of electron–electron interactions [36–39]. High values of magnetoresistance were observed on these systems, ranging from 20% at room temperature up to 50% at 100 K (not shown), indicating effective spin dependent tunneling through the barriers together with cotunneling processes. One should emphasize that the data at liquid helium temperature is expected to be in the 10 pA range, making the use of a vertical geometry [23] or HAR gaps design critical for characterizing the samples at low temperatures, and in this peculiar case, for demonstrating the spintronic properties of nanoscale entities.

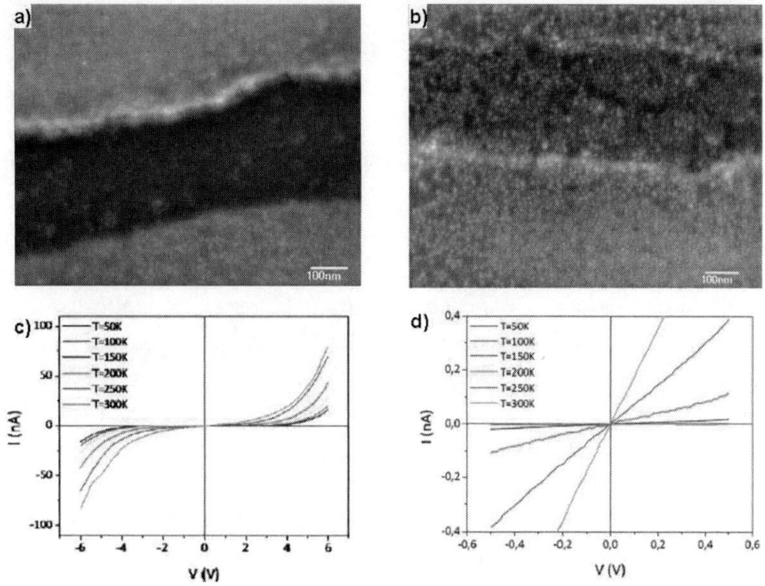

Figure 6.4 Iron oxide nanoparticles over a nanotrench. (a) SEM image of a sample with a single monolayer, (b) with 5 layers. (c) Sub-pA leakage current systematically measured on empty nanotrenches. d) Non-linear $I(V)$ behavior at several temperatures of nanotrench containing Fe_3O_4 nanoparticle multilayers. (d) Zoom-in of the Coulomb-gap part. See also Color Insert.

6.4.2 High-Current Devices

The second device demonstrates the adequacy of nanotrench geometry for particle- mediated multiscale approach. This approach has been proposed recently for studying molecular junctions, where metal-coated micron-sized spheres were magnetically trapped between electrodes patterned by e-beam lithography, covered with a self-assembled monolayer of an organic molecule [40]. Here, we describe a complementary technique which takes advantage of the large length and small width of the nanotrench to allow controlled positioning of an array of Ni/Au coated silica microspheres over the gap by magnetic trapping. Trench geometry is also found to be convenient for direct optical detection of trapping events, thanks to the high contrast of the trench region; providing a simple platform for detecting occurrences of an adequately positioned sphere.

The device fabrication process can be divided into three steps. First, nanotrench electrodes made of a ferromagnetic/metallic bilayer are patterned, followed by the magnetization of ferromagnetic layers using an external magnetic field. In the following example, nanotrenches were made following the previously described process, by using the parameters as followed: first layer deposition of Ti (5 nm)/ Ni (35 nm)/Au (20 nm), second layer deposition at 65° tilt angle of Ti (10 nm)/ Ni (50 nm)/Au (30).

Conductive and magnetically active silica microspheres are then fabricated by depositing a Ti (5 nm)/Ni (30 nm)/Au (20 nm) layer on a hemisphere (Fig. 6.5b). A dispersion of metal-coated microspheres in an organic solvent is locally deposited over the nanotrenches by drop casting (Fig. 6.5c). As a result of the magnetic-field gradient between the nanotrench electrodes, microspheres are trapped over the trench. The process is performed under optical microscope observation where occurrence of microsphere bridging can be immediately assessed. The used off magnetic trap allows one to contact more microspheres in between the electrodes, compared to what can be obtained (cf. Fig. 6.5d,e for comparison) with microspheres coated only with gold [26].

Once spheres are trapped over the gap, the resistance decreases from the open-circuit value to several hundred ohms showing an ohmic IV curve (cf. Fig. 6.5f). Using a pseudo four-point geometry, with patterned y-shaped electrodes for separate current and voltage probes, we can reproducibly observe a sample resistance below 50 Ω at room temperature. Such low-value ohmic contact reveals that stable electrical connections can be made over the trenches. It also confirms the quality of the surfaces and their adequacy for making good electrical interconnects, which is challenging when inserting a 'wet' deposition step in the device build up or for free-standing devices. It also demonstrates the ability of the nanotrenches to sustain high-current density. This multiscale approach can be used for organic and molecular electronics purposes by functionalizing the spheres and/or the electrodes, prior to microsphere deposition over the trench. This strategy is versatile and could be compatible with different sized particles from nm to μm range.

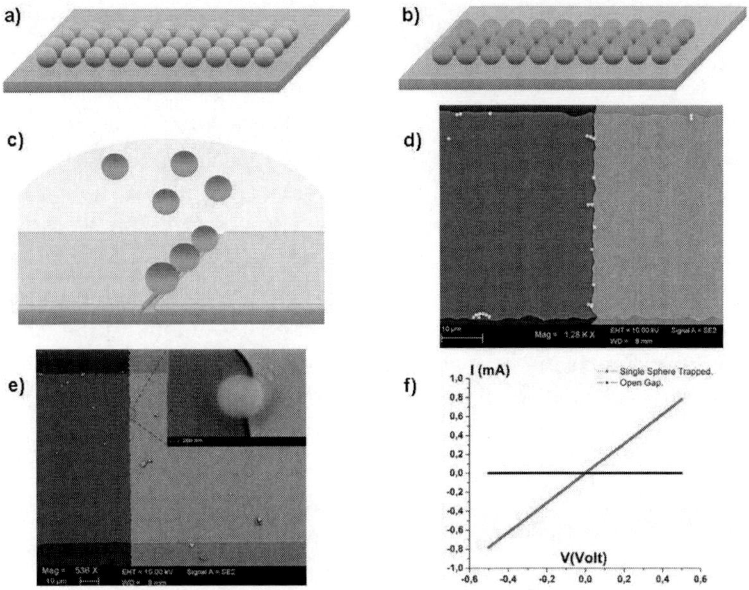

Figure 6.5 Trapping of metal-coated microspheres over a nanotrench. (a) Silica microspheres deposited on glass surface. (b) Metal coating of the microspheres using e-beam evaporation. (c) deposition of microspheres in solution over the nanotrench. (d) SEM image of a 40 μm-long and 100 nm-wide nanotrench with nickel/gold-coated microspheres (diameter = 300 nm) magnetically trapped between the Nickel/Gold electrodes, metal side facing downwards. (e) SEM image of a 100 μm long and 80 nm wide nanotrench with a single gold-only covered microsphere placed between the electrodes. Inset: Higher magnification image showing the microsphere contacts both sides of the gap. The blurred edges are due to charging. (f) $I(V)$ measurements of a nanotrench before (black) and after (red) trapping a single gold-only coated microsphere. See also Color Insert.

6.5 Conclusion

In summary, we showed that several micrometers wide electrodes separated by a sub-100 nm gap can be reproducibly produced by means of simple optical lithography techniques, with a control of the electrodes distance of the order of 10 nm accuracy. The process is simple and easily scalable for high throughput fabrication as it uses standard optical lithography techniques and allows parallel processing of many junctions. The nanotrenches are suitable for use in micro and nano electronics where averaging measurements over several particles contacted in parallel is important and are well adapted to nano and microparticle interconnect based strategies [26]. The device functionality was demonstrated by measuring I-V transport properties of mono and multi layers of highly resistive ferromagnetic nanoparticles from room temperature down to 50 K. We observed manifestation of spin-dependant tunneling in the presence of electron–electron interactions, demonstrating that nanotrenches are a very effective design for measuring transport properties of such highly resistive nanoscale assemblies and for demonstrating their spintronic properties.

Another example of the interest of such a device was demonstrated by their use as a trap for electrically conductive and magnetically susceptible microspheres. The good ohmic contact between the spheres and the electrodes on each side of the nanotrench and the increased probability of trapping spheres between electrodes confirmed the reliability and utility of such kind of electrodes. Both these systems demonstrate that resistance values ranging from 10^2 Ω up to 10^{12} Ω can be reliably measured, with corresponding current from mA down to a few tenth of pA. Owing to their peculiar geometry, the nanotrench electrodes are ideally suited when high dynamic range of current values is being measured. This is especially useful for temperature-dependent studies on samples where thermally activated transport process should result in exponential change of resistance versus temperature.

Materials with large intrinsic sheet resistance values also strongly benefit from large aspect ratios of the interconnects. It is interesting to note that the nature of the deposited material is not important for this process as only the geometrical height of the electrodes is involved in the shadowing effect. Manufacturing gaps with two different electrode materials is, therefore, straightforward, in contrast with e-beam patterning strategies. This can be of particular interest when designing diodes or photo-active particles combined with molecular electronic systems. The process can also be straightforwardly extended to fabricate magnetic electrodes, of interest for spintronic applications, or to fabricate superconducting electrodes, relevant for mesoscopic physics and quantum effects.

Acknowledgments

Work was performed at the Strasbourg Nanofabrication Facility (STnano). We thank the clean-room staff Dr. H. Majjad, and S. Siegwald for their help, and F. Chevrier for his cryogenic technical support. This work was supported by the Agence Nationale pour la Recherche (ANR Blanc MOSE and ANR MAGARRAYS), European Nano SciERA (INTERNET) grant and the Région Alsace. We thank the group of Pr. S. Begin for collaboration and accepting publishing data related to the Fe_3O_4 nanoparticles part.

References

1. For a book, see, for example, J.M. Tour, *Molecular Electronics: Commercial Insights, Chemistry, Devices, Architecture And Programming*, World Scientific, River Edge, NJ, USA, 2003.
2. For a review, see, for example, K.S. Kwok, J.C. Ellenbogen, *Mater. Today*, 5, 28, 2002.
3. D.C. Ralph, C.T. Black, M. Tinkham, *Phys. Rev. Lett.*, 78, 4087–4090, 1997.
4. D.L. Klein, R. Roth, A.K.L. Lim, A.P. Alivisatos, P.L. McEuen, *Nature*, 389, 699–701, 1997.
5. M.A. Reed, C. Zhou, C.J. Muller, T.P. Burgin, J.M. Tour, *Science*, 278, 252–254, 1997.
6. S.J. Tans, M.H. Devoret, H.J. Dai, A. Thess, R.E. Smalley, L.J. Geerligs, C. Dekker, *Nature*, 386, 474–477, 1997.
7. S.M. Wu, M.T. Gonzalez, R. Huber, S. Grunder, M. Mayor, C. Schonenberger, M. Calame, *Nat. Nanotechnol.*, 3, 569–574, 2008.
8. Y. Selzer, L.T. Cai, M.A. Cabassi, Y.X. Yao, J.M. Tour, T.S. Mayer, D.L. Allara, *Nano Lett.*, 5, 61–65, 2005.
9. A. Salomon, D. Cahen, S. Lindsay, J. Tomfohr, V.B. Engelkes, C.D. Frisbie, *Adv. Mater.*, 15, 1881–1890, 2003.
10. M. Pierre, R. Wacquez, X. Jehl, M. Sanquer, Vinet O. Cueto, *Nat. Nanotechnol.*, 5, 133–137, 2010.
11. H.J. Dai, E.W. Wong, C.M. Lieber, *Science*, 272, 523, 1996.
12. J. Liao, L. Bernard, M. Langer, C. Schönenberger, M. Calame, *Adv. Mater.*, 18, 2803, 2006.
13. S. Kinge, M. Crego-Calama, D.N. Reinhoudt, *Chem. Phys. Chem.*, 9, 20–42, 2008.

14. J.M.B. Kyle, E.W. Christopher, S. Siowling, A.G. Bartosz, *Small*, 5, 1600–1630, 2009.

15. M. Pauly, B.P. Pichon, A. Demortière, J. Delahaye, C. Leuvrey, G. Pourroy, S. Bégin-Colin, *Superlattices Microstruct.*, 46, 195–204, 2009.

16. H.B. Akkerman, P.W.M. Blom, D.M. de Leeuw, B. de Boer, *Nature*, 441, 69–72, 2006.

17. S.M. Luber, F. Zhang, S. Lingitz, A.G. Hansen, F. Scheliga, E. Thorn-Csanyi, M. Bichler, M. Tornow, *Small*, 3, 285–289, 2007.

18. C.T. Black, C.B. Murray, R.L. Sandstrom, S. Sun, *Science*, 290, 1131–1134, 2000.

19. S. van der Molen, J. Liao, T. Kudernac, J. Agustsson, L. Bernard, M. Calame, B. van Wees, B. Ferringa, C. Schönenberger, *Nano Lett.*, 9, 76–80, 2009.

20. D. Greshnykh, A. Fromsdorf, H. Weller, C. Klinke, *Nano Lett.*, 9, 473–478, 2009.

21. C. Zhou, M.R. Deshpande, M.A. Reed, L. Jones, J.M. Tour, *Appl. Phys. Lett.*, 71, 611–613, 1997.

22. D.Q. Li, A. Bishop, Y. Gim, X.B. Shi, M.R. Fitzsimmons, Q.X. Jia, *Appl. Phys. Lett*, 73, 2645–2647, 1998.

23. H. Zeng, C.T. Black, R.L. Sandstrom, P.M. Rice, C.B. Murray, S.H. Sun, *Phys. Rev. B*, 73, 020402, 2006.

24. B. de Boer, M.M. Frank, Y.J. Chabal, W.R. Jiang, E. Garfunkel, Z. Bao, Langmuir, 20, 1539–1542, 2004.

25. B.C. Haynie, A.V. Walker, T.B. Tighe, D.L. Allara, N. Winograd, *Appl. Surf. Sci.*, 203, 433–436, 2003.

26. J.-F. Dayen, V. Faramarzi, M. Pauly, N.T. Kemp, M. Barbero, B.P. Pichon, H. Majjad, S. Begin-Colin, B. Doudin, *Nanotechnology*, 21, 335303, 2010.

27. J. Tang, Y. Wang, C. Nuckolls, S.J. Wind, *J. Vac. Sci. Technol. B*, 24, 3227–3229, 2006.

28. A. Fursina, S. Lee, R.G.S. Sofin, I.V. Shvets, D. Natelson, *Appl. Phys. Lett.*, 92, 113102, 2008.

29. R. Negishi, T. Hasegawa, K. Terabe, M. Aono, T. Ebihara, H. Tanaka T. Ogawa, *Appl. Phys. Lett.*, 88, 223111, 2006.

30. J.G. Bai, C.-L. Chang, J.-H. Chung, K.-H. Lee, *Nanotechnology*, 18, 405307, 2007.

31. L.F. Sun, S.N. Chin, E. Marx, K.S. Curtis, N.C. Greenham C.J.B. Ford, *Nanotechnology*, 16, 631, 2005.

32. A. Javey, P. Qi, Q. Wang, H. Dai, *PNAS*, 101, 13408–13410, 2004.

33. J.G. Bai, W.-H. Yeo, J.-H. Chung, *Lab on Chip*, 9, 449–455, 2009.
34. P. Poddar, T. Fried, G. Markovich, *Phys. Rev. B*, 65, 172405, 2002.
35. C.T. Black, C.B. Murray, R.L. Sandstrom, S. Sun, *Science*, 290, 1131–1134, 2000.
36. M. Pauly, J.F. Dayen, D.S. Golubev, J.-B. Beaufrand, B.P. Pichon, B. Doudin, S. Bégin-Colin, *Small*, 8, 108, 2012.
37. A. Zabet-Khosousi, A.-A. Dhirani, *Chem. Rev.*, 108, 4072–4124, 2008.
38. K.K.-H. Müller, J. Herrmann, B. Raguse, G. Baxter, T. Reda, *Phys. Rev. B*, 66, 075417, 2002.
39. D.S. Golubev, A.D. Zaikin, *Phys. Lett. A*, 169, 475, 1992.
40. D.P. Long, C.H. Patterson, M.H. Moore, D.S. Seferos, G.C. Bazan, J.G. Kushmerick, *Appl. Phys. Lett.*, 86, 153105, 2005.

Chapter 7

High-Aspect-Ratio Metallic Nanostructures for Transparent Electrodes

Joong-Mok Park,[a,*] Xinyu Liu,[a] Wai Leung,[a] Kristen Constant,[a,b] Alan Russell,[a,b] and Kai-Ming Ho[a,c]

[a] *Ames Laboratory — U.S. DOE, Ames, Iowa 50011, USA*
[b] *Department of Material Science, Iowa State University, Ames, Iowa 50011, USA*
[c] *Department of Physics and Astronomy, Ames, Iowa 50011, USA*
*joongmok@iastate.edu

Metallic nanowire arrays having high optical transmission and electric conductivity show promise for use as transparent electrodes. Transparent electrodes require high transmission of visible light and good electrical conductivity for charge transfer. High-aspect-ratio metallic nanowires for transparent electrode applications can be fabricated by e-beam angular deposition on polymer templates. These polymer templates are made with interference holography and nanoimprinting using a polydimethylsiloxane (PDMS) mold. The details of the fabrication processes including interference holography, micro-transfer molding, nanoimprint, and shadow angle depositions will be discussed.

Vistas in Nanofabrication
Edited by Faiz Rahman

ISBN 978-981-4364-56-0 (Hardcover), 978-981-4364-57-7 (eBook)
www.panstanford.com

7.1 Introduction

Semiconducting and metallic nanowires have many interesting physical, electrical, optical, and chemical properties due to their simple nature and one-dimensional geometry [1]. Their mechanical and electrical properties depend on the average grain size achieved during their growth. Grain growth mechanisms are well-studied for thin films and bulk materials, but there are few studies on nanowires [2,3]. The average grain size of metallic nanowires and thin films is generally smaller than that of bulk but can be increased by thermal annealing. In our work, metallic nanowires of Ti, Au, and Al were fabricated using physical deposition on polymer templates. The dimensions and grain growth of the nanowires were monitored *in situ* during the annealing process by transmission electron microscopy (TEM). Optical and electrical properties of the high-aspect-ratio metallic nanowire arrays were measured. Metallic or metal oxide thin films are currently used as transparent electrodes, but the thicknesses are constrained by the competing requirements of maximizing optical transparency and electrical conductivity. Typically, thicker films have desirable higher electric conductivity but transmission decreases exponentially due to increased radiation absorption. To decouple electric conductivity and optical transparency, tall and thin (high aspect ratio) metallic nanowires were proposed and fabricated [4]. The high-aspect metallic nanowires have optical and electrical properties similar to indium tin oxide (ITO) and have potential for use as transparent electrodes in many practical optical devices.

7.2 Fabrication of Nanowire Using Polymer Templates

Nanowires were made by e-beam assisted physical vapor deposition of metal on polymer templates. The polymer templates were made either with interference lithography or through nanoimprinting.

7.2.1 Polymer Template Fabrication

Photoresist gratings were fabricated with laser interference holography [5,6] and photosensitive materials (photoresist), which selectively react upon exposure to light, typically ultraviolet.

Instead of a photo mask, the interference of coherent beams is used to construct one- or two-dimensional periodic structures. The interference is the superposition of two coherent beams, the sum of the electric fields, as illustrated in Fig. 7.1. The intensity of the interference is the time-averaged value during the exposure and is simplified as

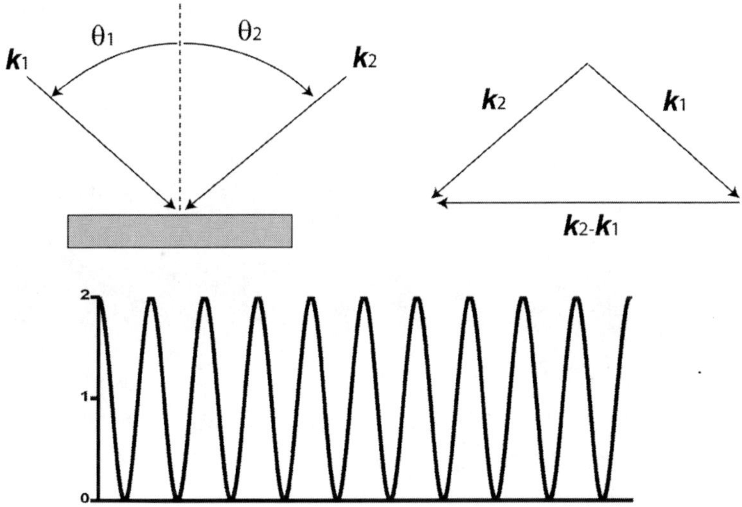

Figure 7.1 Interference of two coherent beams at two incidence angles and the intensity of the interference along the sample surface.

$$I = \langle |E_1 + E_2|^2 \rangle = I_0[1 + \cos(k_2 - k_1)x], \tag{7.1}$$

where E_1 and E_2 are time-dependent electric fields and k_1 and k_2 are wave numbers.

The pitch (P) of the constructive and destructive interference patterns is a function of the laser wavelength (λ) and the angle of incidence (θ) from the normal: $\text{P} = \lambda/(\sin(\theta_2) - \sin(\theta_1))$. Because the intensity of the interference is sinusoidal, a periodic line grating forms in the photoresist. An Ar-ion laser (λ = 364 nm) was used as the coherent light source, and the pitch can be adjusted to $\lambda/2$ by changing the incidence angles. The exposed area is determined by the size of optical components and the coherence length of the laser. Large samples were made by expanding the beams with a UV objective lens and exposed at 2 m from the focal point of the lens. A Lloyd's mirror setup was used as shown in Fig. 7.2a [6].

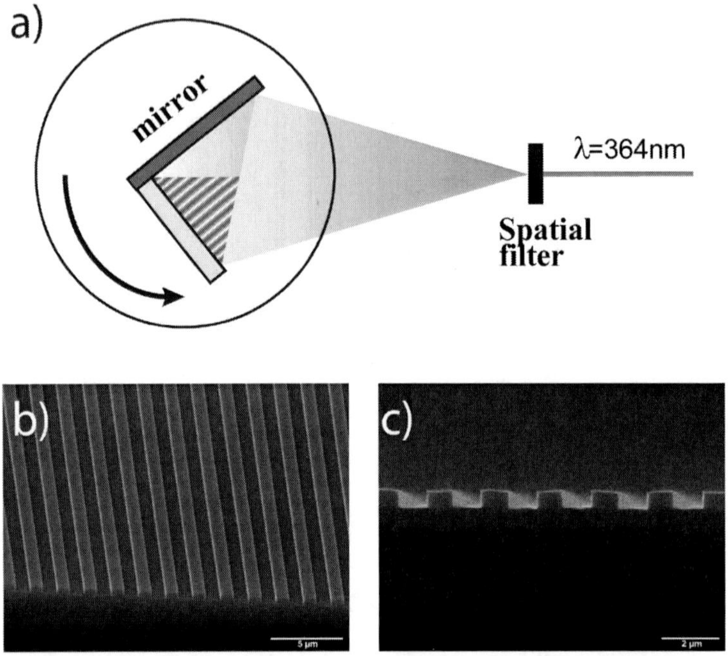

Figure 7.2 (a) Schematic diagram of the interference lithography setup with the Lloyd's mirror configuration. (b) SEM image of photoresist grating made by interference lithography on a silicon substrate. (b) Cross-sectional image of the photoresist.

One reflected beam from a mirror and the other one from the objective lens created the interference pattern. The mirror was attached at right angle to the sample and the assembly was rotated together to maintain the same incidence angles. Following the laser exposure of the photoresist, the sample was developed, leaving a grating structure of photoresist on the substrate as shown in Fig. 7.2b,c. Note that photoresists structures have a rectangular cross section even though the intensity of the light is sinusoidal.

Other well-known methods for making nanostructures with polymer materials are nanoimprinting [7,8] and micro-transfer molding [9]. Here, PDMS is used to replicate the inverse structure of the original master pattern. A second polymer, polyurethane (PU), is used to duplicate the inverse structure of the PDMS mold. The PU then attains the same structure as the original master pattern through double inversion.

7.2.2 Metal Deposition

After polymer templates were made, various metals were deposited on the templates by e-beam evaporation. A deposition rate of 1 Å/s and a pressure below 10^{-6} Torr were maintained during the depositions. Because the distance between the e-beam source and the specimen was 1 m, a distance far enough for line-of-sight deposition, metals were deposited only on the tops and the bottoms of the grating, leaving the sidewall of the grating surfaces uncoated as shown in Fig. 7.3a,b. Thus, two sets of nanowires were produced; one set adhering to the substrate and a second set adhering to the tops of the polymer grating. After deposition, the polymer templates can be dissolved with a photoresist remover (Microchem Inc., Remover PG), releasing the deposited metal nanowires atop the grating into the rinse solution, isopropyl alcohol. The metallic nanowires in solution were then dropped onto a flat substrate and dried in vacuum. The collected nanowires were slightly curled as shown in Fig. 7.4a,b. Half of the metallic nanowires, those that were deposited in the channels between the polymer remained on the substrate even after polymer removal as seen in Fig. 7.4c. Al, Au, and Ti nanowires have smooth surfaces compared to that of Pb nanowires. There was no substrate heating during the deposition except the heat radiation from the source. The high mobility of Pb atoms on the substrate during the initial grain growth stage results in a large grained structure as seen in Fig. 7.4d. For this reason, Pb nanowires cannot be obtained with this method. Additionally, they easily oxidize after a short period of exposure in air.

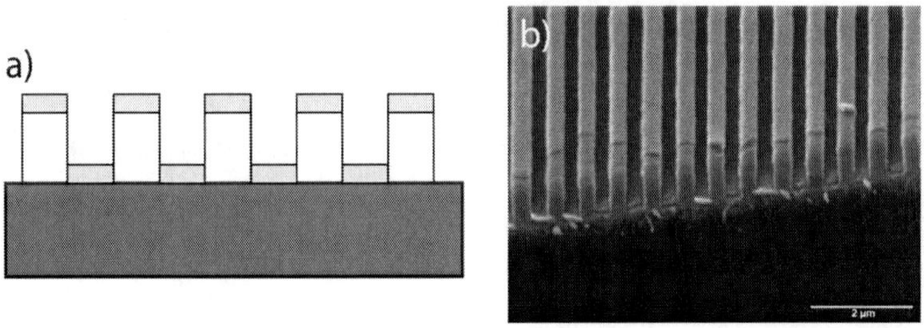

Figure 7.3 (a) Schematic diagram of metal deposition on the polymer template at normal angle. (b) SEM image of 100 nm-thick Al coated on the top and the channels of the photoresist.

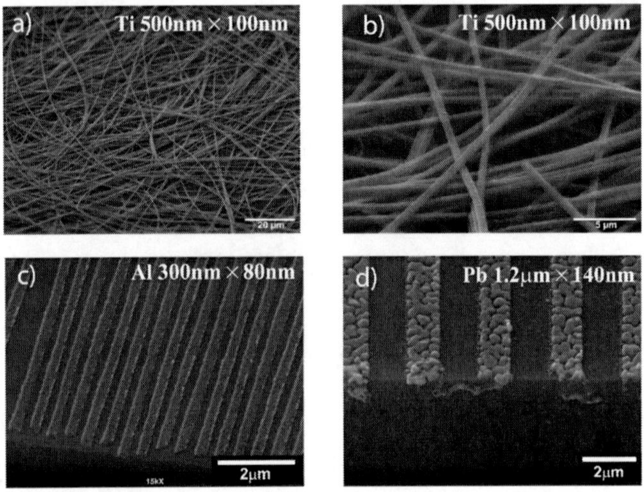

Figure 7.4 (a, b) SEM images of nanowires collected from the solvent used for dissolving photoresist. (c) Al, and (d) Pb nanowires on the substrate after photoresist is removed.

7.3 Characterization of Nanowires

7.3.1 Transmission Electron Microscopy

Once the metal nanowires were made, these were transferred onto a graphite mesh for TEM observations as shown in Fig. 7.5a. The nanowires were broken into lengths of a few microns by ultrasonic agitation for 1 minute in isopropyl alcohol to avoid self-folding. Then the broken nanowires in solvent were dropped onto a graphite mesh and suspended by vacuum-assisted drying. Also, a metal thin film having voids was made to support the ordered metal nanowires instead of using the graphite mesh. Cu thin films were coated by electro-deposition onto the nanowires with the templates on the substrate, and then a hole was selectively etched out of the Cu, leaving metallic nanowires bridging over these holes after separating the Cu thin film from the substrate as shown in Fig. 7.5b. These metallic nanowires were suspended across the gap and adhered to the free-standing Cu substrate on either side of the gap. This allowed the nanowires to span a window in the substrate so an electron beam could pass through the nanowires. In this way the nanowires can be imaged *in situ* by TEM.

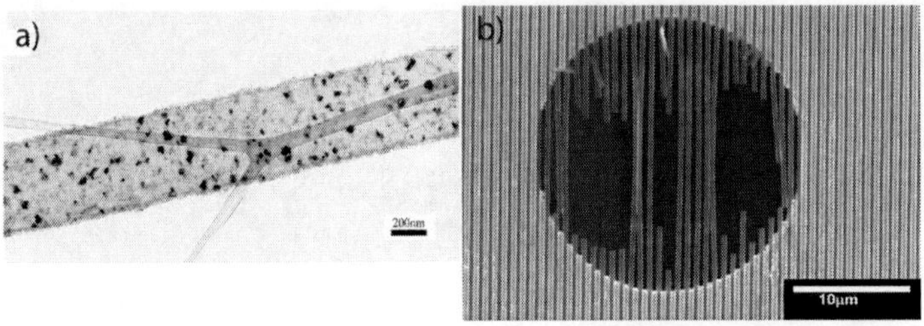

Figure 7.5 (a) TEM micrograph of a 400 nm-wide and 100 nm-thick Al nanowire suspended on a Y-shaped graphite mesh. Some Al grains are oriented so as to appear dark in this image. (b) SEM image of Ti metal nanowires suspended across Cu film having a circular hole.

TEM observation of deposited nanowires shown in Fig. 7.6 shows that the metal is thin enough (~100 nm) to be electron transparent to the 200–300 kV beam energies used in TEM microscopes. The as-sputtered Ti and Al nanowires shown in Fig. 7.6a,c have undesirably small average grain sizes. Such small grains are typical of sputtered films. In addition, dislocation generation and multiplication would be facilitated by larger grains. The average grain sizes were about a few nm for Ti and 10–20 nm for Al nanowires as-grown. Their average grain sizes were increased after annealing, but large or bamboo-like grains were not observed. To check the grain growth over a long period of time, 80 nm-thick Au nanowire samples were placed in an ultra-high vacuum chamber. Au nanowires were annealed for 100 hours at 300°C and at a pressure below 4×10^{-8} Torr. Their average grain size is close to that annealed at 600°C *in situ* for 14 minutes as seen in Fig. 7.7b–d.

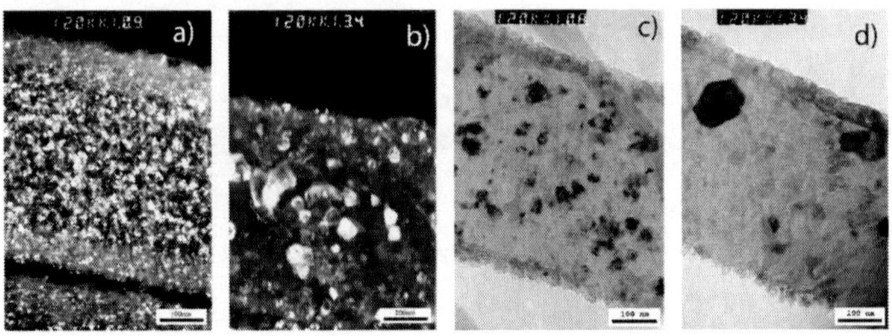

Figure 7.6 TEM micrograph of a 100 nm-thick layer of (a) Ti nanowires before annealing and (b) after annealing at 950°C for 10 minutes. TEM images of (c) Al nanowires before annealing and (d) after annealing at 600°C for 160 minutes.

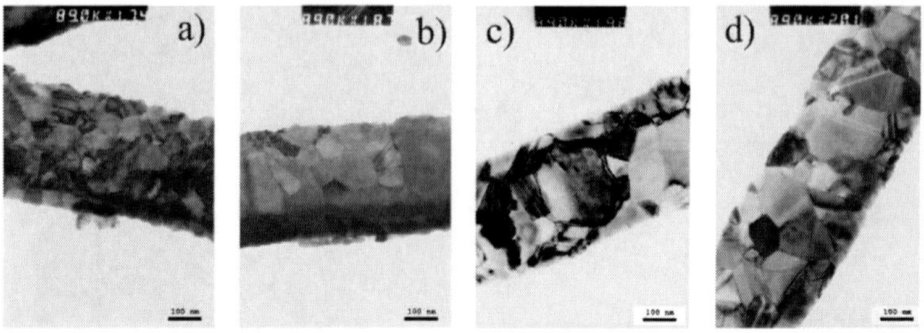

Figure 7.7 TEM micrograph of a 80 nm-thick Au nanowires sample (a) before annealing and (b) after *in situ* annealing at 600°C for 14 minutes. (c, d) Au nanowires after annealing at 300°C for 100 hours under ultra high vacuum. The pressure was below 4 × 10^{-8} Torr during annealing.

7.3.2 *In situ* Grain Growth

Larger grains might be produced by long, high-temperature vacuum anneals to promote grain growth to coarsen the grains of the nanowires. However, previous work on metallic thin films suggests that post-deposition grain coarsening may also result in undesirable effects such as pitting of the surface or otherwise compromising the sample geometry. An alternative approach is to modify metal deposition conditions to control the grain structure. The control of deposition conditions requires modification of the sputtering system to include a rotating, temperature-controlled sample stage, ideally with both heating and cooling capabilities. Figure 7.8 shows TEM images of 100 nm-thick Al nanowires taken during *in situ* annealing at 600°C. A large grain appears abruptly through the merging of two adjacent grains.

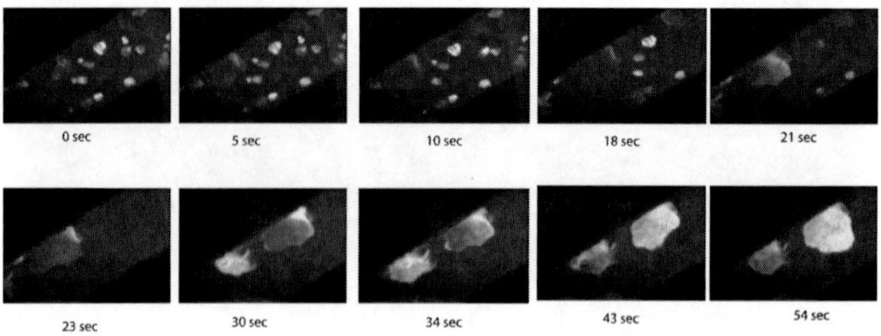

Figure 7.8 TEM micrograph of a 100 nm-thick Al nanowire during *in situ* annealing process at 600°C.

For gold nanowires, we obtained bamboo-like grain structures after *in situ* heating at 700°C for ~1 hour as seen in Fig. 7.9a,b. The diffraction pattern of the electron beam focused on a domain confirmed the existence of the single grain as shown in Fig. 7.9c. This method provided us an opportunity to study the dislocation behavior and allowed us to control the electrical and mechanical properties of nanowires by manipulating the grain structure and grain boundaries.

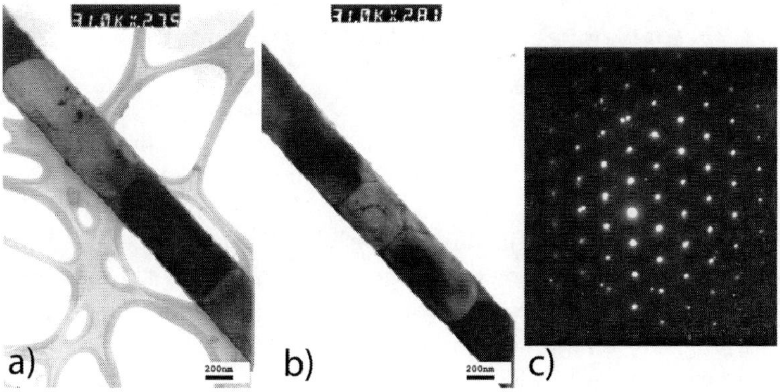

Figure 7.9 (a, b) TEM images of gold nanowires after *in situ* heating at 700°C for ~1 hour. The bamboo-like grain structure is evident. (c) Electron diffraction pattern showing a single grain.

7.4 Shadow Angle Deposition

7.4.1 Shadow Deposition of Metal Nanowires

Figure 7.10a shows a schematic diagram of a shadow-angle-deposition technique [10–12]. The polymer is used as the deposition template and also as a shadow mask. The thickness of the metal on the top and sidewalls is determined by the height of the polymer and the incidence angle. The top thickness is $t\cos(\theta)$, and the sidewall thickness is $t\sin(\theta)$, where t is the normal deposition thickness. For low angles, such as 15°, as in Fig. 7.10b, the top metal coating is thick, while the sidewall is thin. With a 45° angle deposition, the top and sidewall have the same thickness as shown in the inset of Fig. 7.10b.

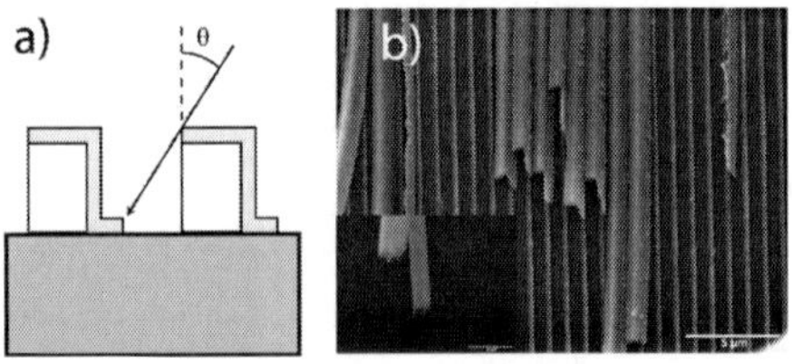

Figure 7.10 Shadow angle deposition and SEM image of Ti nanowires at 15° angle. Inset shows Ti nanowires deposited at 45°.

7.4.2 High-Aspect-Ratio Metallic Structures by Ar Etching

The angular deposition technique can be used to avoid deposition in the polymer channels by shadowing. The metal deposited on top can be removed by physical etching with high-energy Ar ions. The high-energy Ar etches the top, leaving well-aligned high-aspect-ratio metallic structures deposited on the sidewalls as shown in Fig. 7.11a. With controlled etching conditions, the polymer templates can be left intact to provide mechanical support for the high-aspect-ratio nanowires as seen in Fig. 7.11b [4]. Metals were also deposited on both side walls by two depositions at 45° and −45° as seen in Fig. 7.11b. Electron transmission is enhanced by removal of the top metal coating of the polymer, but there is significant diffraction from the polymer grating. This diffraction can be reduced by encapsulation of the polymer with a material having the same refractive index. The electrical conductivity is better than that of ITO, which is often used in displays and photovoltaic applications. This architecture has the potential to be used as a replacement for transparent electrodes without the use of indium, which suffers from high cost and diminishing supplies.

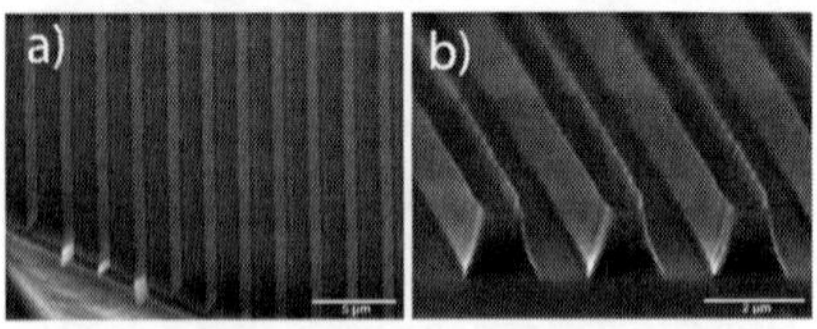

Figure 7.11 (a) After Ar plasma etching the top Au layer deposited on the polymer, and the polymer template were etched away, leaving single sidewalls on the substrate. (b) Etched top of the gold having double sidewalls with Ar ion milling. The polymer templates (PU) were partially etched and remained between the metals.

7.4.3 Electrical and Optical Properties

Electrical conductivity measurement of a single nanowire would require very small electrodes [13–16]. However, measuring a large number of nanowires gives more accurate values and does not require nanometer-scale electrodes or actuators. Simply adding millimeter-wide contact pads of Ag epoxy or thin film metal deposition allows the measurement of resistivity. The contact resistance between nanowires and the contact pad is much less than 1 ohm. A simple two-point measurement is sufficient to measure the resistance of nanowires. Samples with two different heights (300 nm and 1400 nm) were fabricated with PU polymer templates with nanoimprinting [4]. A 40 nm-thick Au layer was deposited on both sides of the polymer, and the top was removed by Ar ion milling. The measured resistance was 16 ohm for 1400 nm tall and 40 nm wide Au nanowires array and 52 ohm for 300 nm tall and 40 nm wide Au nanowires array. The length of the nanowires was 2.5 cm and a total of 4×10^4 nanowires measured at the same time. The typical resistance of 150 nm ITO-coated glass having the same dimensions is 30 ohms.

The optical transmission of the grating structures exhibits diffraction orders dependent on the dimensions of the gratings including pitch and height. Such diffraction can be utilized in various applications. The optical transmissions of the high-aspect-ratio metallic nanowires with PU templates were measured at normal incidence angle after the top Au layer was removed. The specular transmission is the transmitted intensity of the 0th order diffracted light, and the total transmission is the integrated transmitted intensity of all diffracted orders. The total transmission was measured by placing the sample on a 3 mm opening of the input port of a 75 mm diameter integrating sphere. A fiber-coupled spectrometer (Ocean optics S2000) analyzed the transmission spectrum. As seen in Fig. 7.12, specular transmissions are increased by encapsulation, which indicates that the higher-order diffractions are suppressed. This is essential for some applications such as smart windows, displays, or back-lighting. Also, there is a slight (few %) enhancement of total transmission after encapsulation. This is due to the over-etched polymer being filled by encapsulated materials.

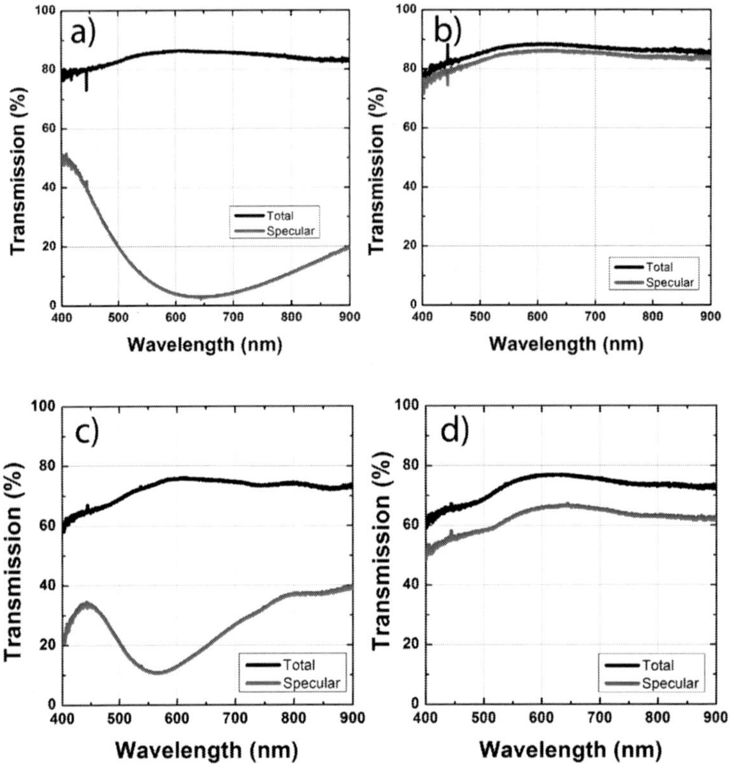

Figure 7.12 Optical transmission of high-aspect-ratio Au nanowire array after removal of the Au deposited on the top of the polymer. (a) The height = 300 nm and width = 40 nm before and (b) after encapsulation. (c) The height = 1400 nm and width = 40 nm before and (d) after encapsulation. See also Color Insert.

7.5 Conclusions

In summary, we fabricated metallic nanowires with laser interference holography, nanoimprint, and e-beam angular metal depositions. This fabrication method can produce well-defined, rectangular cross-sectional nanowires with dimensions of a few cm in length and submicron width. Structural properties of metallic nanowires were examined with SEM and TEM. The high-aspect-ratio metallic nanowires show both high optical transmission and high electrical conductivity after the metals deposited on top were removed. Encapsulation of the nanowires suppressed the high-order diffractions.

Acknowledgments

This work was supported by the Director for Energy Research, Office of Basic Energy Sciences. The Ames Laboratory is operated for the U.S. Department of Energy by Iowa State University under contract No. W-7405-Eng-82.

References

1. Pauzauskie, P.J., and Yang, P. (2006), Nanowire photonics, *Mater. Today*, **9**, pp. 36–45.
2. Thompson, C.V. (1990), Grain growth in thin films, *Annu. Rev. Mater. Sci.*, **20**, pp. 245–268.
3. Fayad, W., Thompson, C.V., and Frost, H.J. (1999), Steady-state grain-size distributions resulting from grain growth in two dimensions, *Scripta Mater.*, **40**, pp. 1199–1204.
4. Kuang, P., Park, J.M., Leung, W., Mahadevapuram, R.C., Nalwa, K.S., Kim, T.G., Chaudhary, S., Ho, K.M., and Constant K. (2011), A new architecture for transparent electrodes: relieving the trade-off between electrical conductivity and optical transmittance. *Adv. Mater.*, 23, 2469–2473.
5. Prenen, A.M., Werf, J.C.A., Bastiaansen, C.W.M., and Broer D.J. (2009), Monodisperse, polymeric nano and microsieves produced with interference holography, *Adv. Mater.*, **21**, pp. 1751–1755.
6. Park, J.M., Nalwa, K.S., Leung, W., Constant, K., Chaudhary, S., and Ho, K.M. (2010), Fabrication of metallic nanowires and nanoribbons using laser interference lithography and shadow lithography, *Nanotechnology*, **21**, 215301 (6pp).
7. Guo, L.J. (2006), Nanoimprint lithography: methods and materials requirements, *Adv. Mater.*, **19**, pp. 495–513.
8. Kang, M.G., and Guo, L.J. (2007), Nanoimprinted semitransparent metal electrodes and their application in organic light-emitting diodes, *Adv. Mater.*, **19**, pp. 1391–1396.
9. Lee, J.H., Kim, C.H., Ho, K.M., and Constant, K. (2005), Two-polymer microtransfer moldings for highly layered microstructures, *Adv. Mater.*, **17**, pp. 2481–2485.
10. Chen, Y., and Goldman, A.M. (2008), A Simple approach to the formation of ultranarrow metal wires, *J. Appl. Phys.*, **103**, 054312 (4pp).

11. Chen, L., Wang, J.J., Walters, F., Deng, X., Buonanno, M., Tai, S., and Liu, X. (2007), Large flexible nanowire grid visible polarizer made by nanoimprint lithography, *Appl. Phys. Lett.*, **90**, 063111 (3pp).
12. Bai, J.G., Chang, C.L., Chung, J.H., and Lee, K.H. (2007), Shadow edge lithography for nanoscale patterning and manufacturing, *Nanotechnology*, **18**, 405307 (8pp).
13. Peng, Y., Cullis, T., and Inkson, B. (2008), Accurate electrical testing of individual gold nanowires by *in situ* scanning electron microscope nanomanipulators, *Appl. Phys. Lett.*, **93**, 183112 (3pp).
14. Wu, Y., Xiang, J., Yang, C., Lu, W., and Lieber, C.M. (2004), Single-crystal metallic nanowires and metal/semiconductor nanowire heterostructures, *Nature*, **430**, pp. 61–65.
15. Wu, B., Heidelberg, A., and Boland, J.J. (2005), Mechanical properties of ultra-high strength gold nanowires, *Nat. Mater.*, **4**, pp. 525–529.
16. Xiang, C., Kung, S.C., Taggart, D.K., Yang, F., Thompson, M.A., Guell, A.G., Yang, Y., and Penner, R.M. (2008), Lithographyically patterned nanowire electrodeposition: a method for patterning electrically continuous metal nanowires on dielectrics, *ACS Nano*, **2**, pp. 1939–1949.

Chapter 8

Fabrication of Nanogap Electrodes by Electroless- and Electro Deposition

Luis De Los Santos Valladares,[a,b,*] Angel Bustamante Domínguez,[c] Thanos Mitrelias,[a,d] Crispin H.W. Barnes,[a] J. Albino Aguiar,[e] and Yutaka Majima[b]

[a] *Cavendish Laboratory, University of Cambridge, J.J. Thomson Av., Cambridge CB3 0HE, UK*
[b] *Materials and Structures Laboratory, Tokyo Institute of Technology, 4259 Nagatsuta-cho, Midori-ku, Yokohama 226-8503, Japan*
[c] *Facultad de Ciencias Físicas, Universidad Nacional Mayor de San Marcos, Ap. Postal 14-0149, Lima, Perú*
[d] *Cambridge Biomagnetics Ltd., St. Johns Innovation Centre, Cowley Road, Cambridge CB4 0WS, UK*
[e] *Laboratório de Supercondutividade e Materiais Avançados, Departamento de Física Universidade Federal de Pernambuco 50670-901, Recife, Brazil*
*ld301@cam.ac.uk, luisitodv@yahoo.es

In this chapter, the fabrication of metal nano-spaced electrodes for electronic nanodevices by electro and electroless plating is discussed. The necessary reagents, conditions, and processes required to obtain nano and atomic gaps between soft and clean surfaces electrodes are described. In the electroless method, the plating process is performed catalytically after immersing the sample in a solution which contains the same metal ions. In the electrodeposition technique, metal ions of an electrolyte move towards the sample under an applied voltage.

Vistas in Nanofabrication
Edited by Faiz Rahman

ISBN 978-981-4364-56-0 (Hardcover), 978-981-4364-57-7 (eBook)
www.panstanford.com

Both techniques are explained with examples, the first technique is described by demonstrating the formation of gold nanogap electrodes using common medical solutions as reactants, whereas the second technique is described by showing electrodeposition of nickel electrodes in a conventional electrochemical cell. Current voltage characteristics are also presented to evaluate possible applications of the nanogap electrodes in electronic nanodevices.

8.1 Introduction

Nano and molecular electronics devices require the fabrication of symmetric metal electrodes separated by a nanogap ("nanogap electrodes") in which a specific molecule or crystal can be placed in order to connect them to the macroscopic world. In the last two decades, vertical structures in which a self-assembled monolayer (SAM) of molecules is electrically connected on one side with a scanning tunneling microscope (STM) [1,2] or conductive probe atomic force microscope (C-AFM) [3] and on the other side by a metallic surface have been demonstrated. Even though this approach has yielded many important results, it suffers from limitations such as the enormous asymmetry of the electrodes, the requirement of high vacuum environment, difficulties in mass production, and difficulty in maintaining a stable chemical bond between the molecule and the microscope tip due to mechanical vibrations. To solve these problems, more recently, new coplanar metal/molecule/metal devices have been proposed. Nanogap electrodes are fabricated before the molecular components, and they are subsequently inserted. This methodology has the advantage that the junction can be characterized with and without the presence of the molecule, thus, allowing the characterization of the molecule. Among others, there are three most remarkable new approaches for making in-plane nanogaps: (i) controlling a break junction mechanically, (ii) electrical breakdown of thin metal wire via electromigration, and (iii) electroless and electrochemical plating. These techniques are schematically represented in Fig. 8.1. The first technique was first developed by Moreland and Ekin [4]. An Nb–Sn wire mounted on a flexible glass beam can be broken to form an electron tunneling junction between the fracture elements. The method was later improved by other researchers. Notched wires of different metals are obtained first with lithography, by bending the substrate with a piezoelectric transducer connected to a pushing rod, and the gap is then formed after breaking the neck (Fig. 8.1a) [5–12]. Using this method, the gap between the electrodes can be flexibly and precisely adjusted to fit different molecules, however, it cannot be removed from the apparatus and, thus, is unsuitable for solid-state device applications. In contrast, in the electromigration-induced break junction technique (EIBJ technique, Fig. 8.1b), the nanogap is produced by applying large current densities to the wire. At high current densities, momentum transfer from the electrons to the ions in the lattice leads to the gradual drift of some ions in the direction of electron flow [13–21]. This mass flux can lead to the growth of voids in the wire, finally causing the formation of the gap [15]. This process is useful for obtaining nanogaps up to 1 nm in length with high efficiency [19], and it can be controlled by observing the Current–voltage characteristics until a tunneling signal is sensed [19, 22]. However, to date, large-scale fabrication of multiple nanogaps in an array of electrodes is still difficult to achieve by this technique. In the

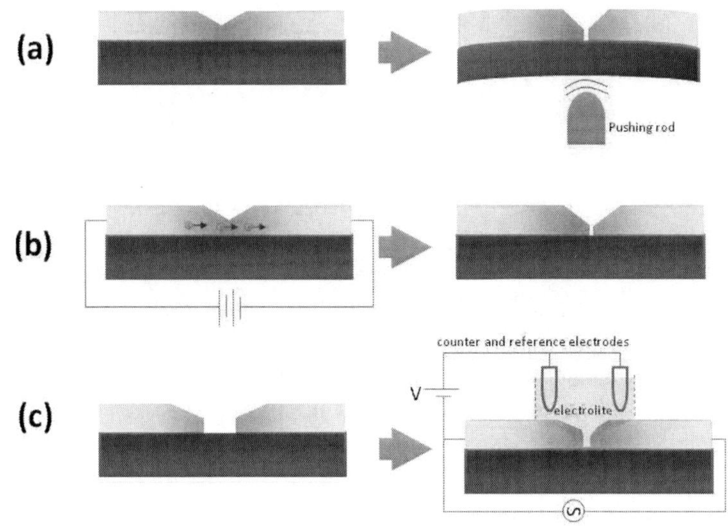

Figure 8.1 In-plane techniques for the fabrication of nanogaps: in the break junction technique (a) the gap is formed after bending the flexible substrate; in the EIBJ technique (b) high-density currents allow the electromigration of atoms to form the gap; and in the electroless and electrochemical technique (c) electrodes forming an initial wide gap are plated to reduce the gap separation.

chemical and electrochemical technique, common lithographically obtained electrodes are grown to reduce their initial separation. The process takes place by submerging the electrodes in an electrolyte usually containing different types of metal ions (Fig. 8.1c). The metal ions can be selectively deposited onto the electrodes surface, assisted by reductive agents. The gap distance depends on the reaction time and reactant concentration (electroless deposition). The plating can be improved by applying a bias voltage between the electrodes (electrodeposition) [23–32]. Furthermore, by using a feedback system, the process can be reversed in the dissolution mode for controlled etching of atoms from the electrodes surface to the solution, and, thus, the gap dimension can be monitored. To date, the electroless and electrochemical plating techniques have been demonstrated to be the most successful techniques for the fabrication of metal electrodes separated by a nanogap. Next, we describe the electroless- and electrodeposition techniques to obtain nanogap electrodes giving one example for each case.

8.2 Nanogap Electrodes Obtained by Electroless Deposition

Electroless deposition is a well-known technique for coating different surfaces with metal films. It is also known as non-galvanic or auto-catalytic plating. There are plenty of literature reports about the application of this technique on relatively large area surfaces [33]. A solid surface (metals for the cases of our interest) is exposed I order to interact with an electrolyte containing metal ions. When interacting, both oxidation and reduction (redox) reactions occur. The oxidation of a substance is characterized by the loss of electrons, while the reduction is distinguished by a gain of electrons. Metal displacement from the salt solution to the metal surface is achieved by reduction, and thus a catalytic component in the solution is required. Silver, nickel, copper, and gold are the most common metals used to fabricate thin metal films by this technique. Among these, gold is nowadays the most commonly deposited metal by this technique and used in nanoelectronics. Gold is favorable because it is highly conductive, it is not easily oxidized, and it allows the formation of self-assembled monolayers of functional molecules such as amines, sulfides, and alkane-thiols on its surface. For instance, alkane-thiol molecules are nowadays used in prototypes of single-electron transistors (SET) to bridge gold electrodes and different type of nanoparticles [34–36]. The fabrication of nanogap electrodes by chemical deposition requires initial metal electrodes which can be obtained by conventional lithography. In the next section, we describe the electroless plating technique taking the deposition of gold electrodes as an example.

8.2.1 Fabrication of Gold Nanogaps

During earlier years, electroless plating of gold has been usually performed by using Au(I) cyanide complex as the gold source. During plating, chemical reactions release cyanide ions CN^- which are toxic [37]. Nowadays, cyanide-free type plating bath is strongly preferred due to safety and environmental concerns. For instance, autocatalytic electroless gold plating (EGP) baths using potassium gold cyanide as gold source and alkali metal borohydride or dimetrhylamine borane as reducing agents have been proposed by Okinaka [38]. However, practical stable non-cyanide electroless plating baths which can meet the severe requirements for high performance electronics manufacturing applications are currently under extensive study. Here, we describe a non-toxic route for plating gold using common medical solutions [39]. The plating process is carried out by immersing the initial lithographically fabricated electrodes in a solution containing $[AuI_4]^-$ ions, these ions are catalyzed and bond to the electrodes forming the plated layer (see Fig. 8.2). As long as there are enough plating ions in the electrolyte, this process continues. The growth in the narrow gap space between the electrodes is a consequence of mass transport from the solution to the gap region. The final gap size is controlled by the concentration of gold ions in the electrolyte and the immersion time.

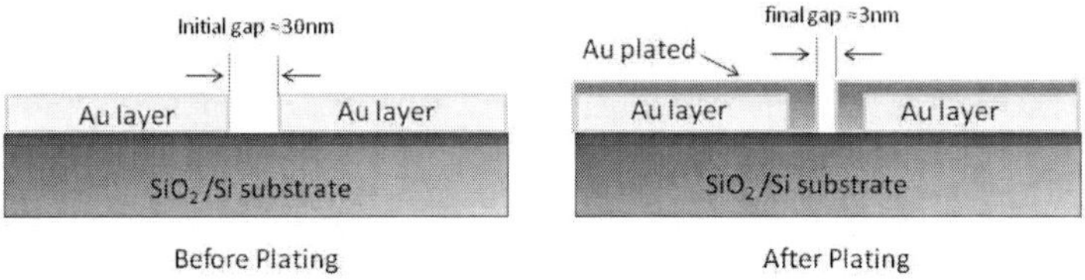

Figure 8.2 Representation of a nanogap between a pair of gold electrodes obtained by electroless deposition of gold. The initial gold layers are obtained by conventional lithography (left side). The gap separation decreases by consecutive plating (right side).

8.2.1.1 The electrolyte

Here, we describe the plating solution that utilizes non-toxic components. The electrolyte is prepared by immersing two thin films of gold (gold sheets of 109 × 109 mm^2, ~150 nm thickness, 99.99% purity, and ~65 mg) in 3mL of commercially diluted iodine tincture consisting of I_2 and KI in ethanol. The chemical formula of the tincture is

$$2KI + I_2 \Leftrightarrow 2K^+ + I_3^- + I^- \tag{8.1}$$

The presence of the tri-iodide ion (I_3) makes the electrolyte dark in color. When the gold sheets are added to the iodine tincture, the tri-iodide ion etches and dissolves the gold sheets to become gold clusters in the solution. This process is facilitated by mixing the components (e.g., by ultrasonic bath for 3 hours), $[AuI_4]^-$ is obtained through the equation

$$2Au + 3I_3^- \Leftrightarrow 2[AuI_4]^- + I^- \tag{8.2}$$

During plating, the tri-iodide needs to be eliminated to avoid dissolution of the gold electrodes; this is achieved by adding $L(+)$ ascorbic acid to the solution. This acid can be easily obtained from commercial Vitamin C. The $L(+)$ ascorbic acid oxidizes I_3^- to I^-.

$$I_2 + I^- \Leftrightarrow I_3^- \tag{8.3}$$

Reduction of I_3^- changes the color of the solution from dark to golden transparent.

To eliminate the remaining gold clusters from the electrolyte and prevent the physisorption of clusters on the electrode surfaces during plating which could block the nanogap, the electrolyte is heated near the melting point of the $L(+)$ ascorbic acid (~190°C) for a few seconds. This allows the tri-iodide ions to further dissolve the remaining gold clusters suspended in the solution.

The electrolyte is dissolved in H_2O to form the EGP solution. The concentration rate EGP/H_2O is one of the factors that need to be taken into account to control the plating. We have observed that the concentration of 1/1000 allows control of the plating process with time. Higher concentrations result in uncontrollable plating speed and physisorption of gold clusters, whereas lower concentrations result in slow plating and in some cases etching of the electrodes due to the generation of tri-iodide ions.

8.2.1.2 Nanogap formation

The electroless plating process takes place by allowing the interaction of the gold electrodes with the EGP solution. This can be easily performed by immersing the clean electrodes in a container with the solution. Due to the interaction of the solution with the electrodes, surface catalysis with the $[AuI_4]^-$ ions occur. The catalysis stops when the amount of gold ions in the solution is exhausted. Figure 8.3 shows nanogap formation in a four electrodes configuration for a molecular transistor in which the horizontal electrodes were designed for later attachment of a molecule, and the vertical electrodes played the role of in-plane gates. With the concentration of the EGP solution 1/1000 and the dimensions of the electrodes shown in Fig. 8.3a, we required five baths of 2 min each to obtain a gap less than 3 nm. During the plating process, the electrodes grow; the probability of the gold ions from the EGP solution to access the nanogap is lower than in larger surface areas, and, thus, the deposition rate in the nanogap region decreases. The process can be stopped when the nanogap is formed (Fig. 8.3b).

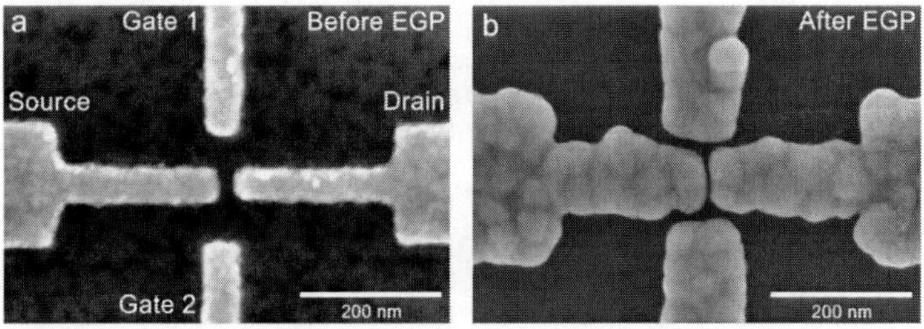

Figure 8.3 Nanogaps electrodes of gold (a) before electroless plating (~30 nm width), initial electrodes can be obtained by conventional lithography. (b) after electroless plating (~3 nm width).

It is important to mention here that in contrast to the break junction technique and the EIBJ technique, the electroless plating technique allows large-scale nanogap production. Multiple nanogaps can be obtained at the same time by depositing an array of multiple initial electrodes. Up to 40% yield of sub 5 nm nanogaps can be obtained from an array of initial gaps of ~30 nm in one process. Furthermore, molecular transistor configurations, in which in-plane gate electrodes are required to be close to the nanogap can be obtained by approaching the gate electrodes with the nanogap electrodes through the EGP process.

8.2.2 Characterization

Single-electron devices consisting of octanethiol (C8S)-protected Au nanoparticle were fabricated by using nanogap electrodes made by the EGP process described above [39]. Figure 8.4 shows the current–voltage characteristics of the gold nanogap before and after Au nanoparticle/chloroform solution dispersion at room temperature. After Au nanoparticle deposition, we observed an increase in the nanogap conductance due to the capture of C8S-protected Au nanoparticle in the gap. After confirming that the C8S-protected Au nanoparticle was trapped between the gold electrodes, we measured the Current–voltage characteristic of the C8S-protected Au nanoparticle at liquid nitrogen temperature in vacuum. As shown in Fig. 8.4, a Coulomb staircase was clearly observed, and the Coulomb gap was estimated to be approximately 0.28 V. By applying the full orthodox theory of single electron charging in double-barrier tunneling junctions, we evaluated each of the two capacitances (C_1, C_2) between the C8S-protected Au nanoparticle and the gold electrodes to be approximately 0.58 and 0.59 aF, respectively. We also evaluated the two tunneling resistances (R_1, R_2) between the C8S-protected Au nanoparticle and the gold electrodes to be approximately 350 and 320 MΩ, respectively. Since the junction parameters of the two junction capacitances and tunneling resistances were almost the same, the structure of the device was expected to be symmetric. These junction parameters were in good agreement with those mentioned in previous reports.

A SET was also fabricated using EGP nanogap electrodes and a chemisorbed Au nanoparticle [40]. Figure 8.5 shows the experimental differential conductance (dI/dV_d) plot in the gate voltage (V_g)–drain voltage (V_d) plane. Coulomb diamonds were clearly observed on SETs fabricated by bottom-up processes of EGP nanogap electrodes and chemisorption of a Au nanoparticle at 80 K. In the drain current–drain voltage characteristics, Coulomb staircases were modulated by the side gate voltage. Tunneling resistances and source/drain/gate capacitances of the SET were evaluated by fitting the theoretical Coulomb staircase determined on the basis of the full orthodox theory in a double-barrier tunneling junction, to the experimental results of

Coulomb blockade under the application of side gate voltages. The theoretical results for the Coulomb diamond are in good agreement with experimental results.

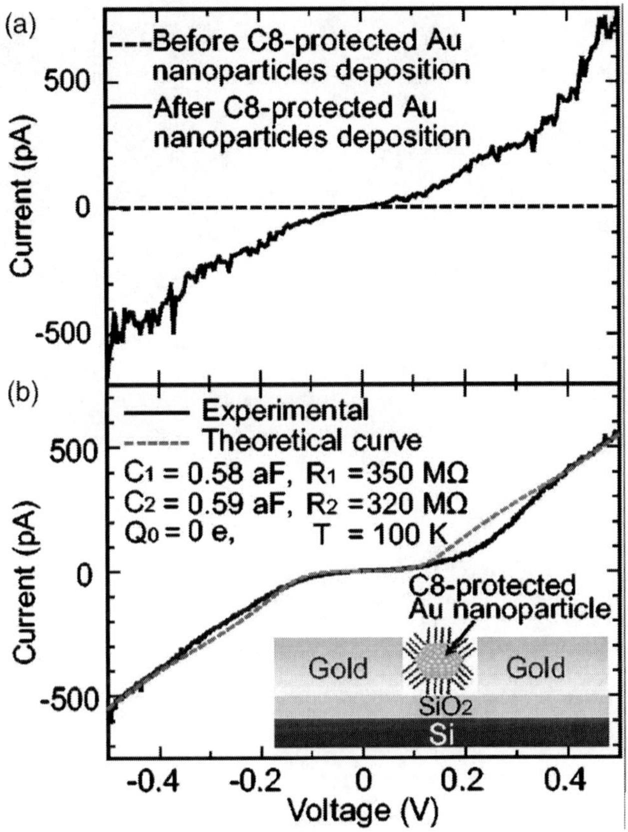

Figure 8.4 Current–voltage characteristics of nanogap (a) before dispersion of C8S-protected Au nanoparticle/chloroform solution (dashed line) and after dispersion of C8S-protected Au nanoparticle/chloroform solution (solid line) at room temperature. (b) Current–voltage characteristics of double-barrier tunneling junction, which consist of gold electrode/C8S-protected Au nanoparticle/gold electrode measured at liquid nitrogen temperature in vacuum (solid line) and the theoretical fitting curve (dashed line). The inset in (b) shows the schematic image of the fabricated device. Nanogap electrodes of gold. (a) before electroless plating (~30 nm in separation), initial electrodes can be obtained by conventional lithography. (b) after electroless plating (~3 nm in separation).

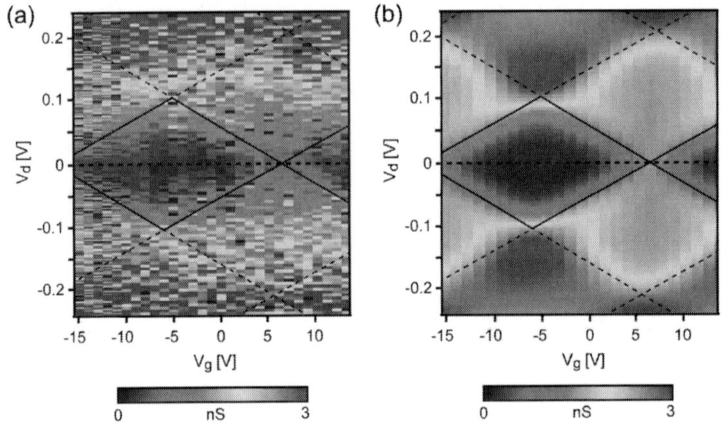

Figure 8.5 (a) Experimental differential conductance dI/dV_d plots in V_g–V_d plane in SET structure consisting of Au drain electrode/$C_{10}S_2$ molecule/Au core/$C_{10}S_2$ molecule/source electrode and side-gate electrode at liquid nitrogen temperature under vacuum. (b) Theoretical dI/dV_d plots in V_g–V_d plane calculated using the following parameters: R_1 = 170 MΩ, R_2 = 160 MΩ, C_1 = 0.76 aF, C_2 = 0.71 aF, and T = 80 K. In both figures, the conductance peak lines calculated from the theoretical fitting under a low temperature condition of T = 10 K are shown as solid and dashed lines. See also Color Insert.

8.3 Nanogap Electrodes Obtained by Electrodeposition

The chemical plating process can be improved by directing the metal ions from the electrolyte to the electrode's surface; this is achieved by creating a potential difference between the metal surface and an auxiliary electrode. This is a well-known process called electrodeposition, and it usually takes place in a chemical cell [37]. Figure 8.6 shows a typical electrochemical cell. The potential is controlled and measured between a reference electrode (RE) and the working electrode (WE-cathode) in which the sample is placed. In addition, an auxiliary electrode (counter electrode, CE) is placed in the anode. The current is measured between cathode and anode. Voltage differences produce redox reaction in the system resulting in the deposition of the metal the ions (M) onto the sample. In the anode, the metal ions are oxidized becoming M^{Z+}. These ions associate with other ion components in the electrolyte and are attracted by the negative charge of the sample surface. After a certain period of time a dynamic equilibrium results.

$$M^{Z+} + Ze \leftrightarrow M \tag{8.4}$$

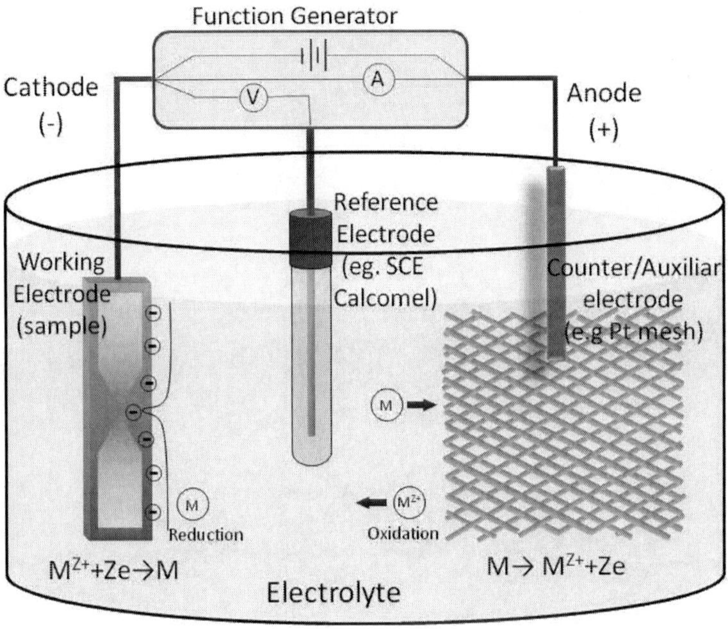

Figure 8.6 Conventional electrochemical cell for electrodeposition. The sample is placed in the cathode and the voltage is referenced to a reference electrode.

Table 8.1 summarizes some of the most noticeable results and conditions for the fabrication of metal nanogaps using the electrochemical technique. When surveying the literature, we found that, with some exceptions, e-beam lithography (EBL) or focused ion beam (FIB) milling techniques are mostly used to obtain the initial electrodes with separation in the range of 40–400 nm. Most authors use lock-in amplifiers to control the deposition and to study the quantum conductance *in situ*. Little consideration has been given to the tunneling current enhancement caused by the electrochemical potential relative to the RE [41] or by the presence of the ions in the electrolyte [25,32,42,43]. Moreover, although most of the experiments listed in the table provide general views about the electrodeposition

Table 8.1 Electroplating conditions and results for the fabrication of nanogaps and nanocontacts reported in the literature

Electrode metal	Deposition metal	Initial electrodes fabrication	Main electrolyte composition	Initial gap width (nm)	Final gap width (nm)	*Ex situ* measurement	Ref.
Gold	Gold	EBL	0.01M $KAu(CN)_2$	50–400	5	NM	[32]
Gold	Gold	EBL	0.1mM KAu	100	<1	*I–V*	[27]
Gold	Gold	EBL	0.01M $KAu(CN)_2$	50–200	10	NM	[31]
Gold	Copper	EBL	0.02M $CuSO_4$	250	2	*I–V*	[23]
Gold	Nickel	FIB	1.5M $Ni(SO_3NH_2)_2$	100	contact	NM	[30]
Gold	Nickel	EBL	0.01–0.02M $NiSO_4$	100	contact	*I–V*	[29]
Gold	Nickel	Nanoporous	0.50M $NiSO_4$	60–120	contact	NM	[47]
Gold	Cobalt	FIB	$CoSO_4$	100	contact	NM	[28]
Gold	Cobalt	EBL	0.22M $CoSO_4$	250	2	*I–V*	[23]
Platinum	Gold	EBL	0.01M $KAu(CN)_2$	50–200	8	NM	[31]
Platinum	Platinum	EBL	0.10M K_2PtCl_4	40–80	20–3.5	*I–V*	[25]
Nickel	Nickel	FIB	H_3NiSO_3	100	contact	NM	[26]

Note: Lock-in amplifiers were used to measure the quantum conductance *in situ*. EBL: Electron (e)-beam lithography; FIB: Focus ion beam milling; NM: No measurement has been performed.

process, they do not provide enough information about the reactions that take place or details on the electrolyte characterization, which are also necessary in order to prevent the formation of passivation or oxidation layers on the surface. It is also seen from the table that gold electrodes are preferred by most authors. In fact, as mentioned above, gold is a good conductor and prevents oxidation. However, magnetic deposition on magnetic metal electrodes can be more attractive since they allow the study of single charge and spin transfer. For instance, ferromagnetic electrodes forming a point contact could be very helpful for studying the controversial ballistic magnetoresistance (BMR) effect [26,44–49]. In the next section, we describe the fabrication of nickel nanogaps by the electrodeposition technique.

8.3.1 Fabrication of Nickel Nanogaps

In this section, we describe the fabrication of nickel nanogap electrodes using a standard electrochemical cell. We characterize the electrolyte, describe the electrochemical conditions, and study the time evolution of the electrodes' growth up to the point of contact. During the plating process, the separation between the electrodes decreases exponentially with time and the nanogap is achieved by stopping the electroplating just before the electrodes touch. In addition, we demonstrate that even if the electrodes are completely bridged during the electroplating technique, the EIBJ technique mentioned in section 8.1 can be successfully applied to re-open the nanogaps.

8.3.1.1 The electrolyte

The electrolyte was prepared by dissolving nickel sulfate ($NiSO_4$, Sigma Aldrich 13635) and boric acid (H_3BO_3, Sigma Aldrich B9645) in water. H_3BO_3 is commonly used in pure nickel and nickel alloys electroplating, because it shows some interesting effects such as excellent buffering to maintain a steady pH in the electrolyte, surface activity, resistance to hydrogen evolution reaction, and improved current efficiency [50–55]. Since we are plating at the nanoscale, the addition of any other additives are not recommended in order to avoid the risk of contamination, passivation layer formation, and rough surfaces. After trying different concentrations of the electrolyte components, we found that the best control of nano-deposition characteristics is achieved by combining 50 mM of $NiSO_4$ and 500 mM of H_3BO_3, whereas for micro-sized or larger area characteristics higher concentration of nickel sulfate is required. We used a standard electrochemical cell for the plating process as represented in Fig. 8.6. The RE consisted of a commercial SCE saturated Calcomel KCl and the C.E was a platinum mesh. A potentiostatic control ensured that the real potential E(t) at the WE with respect to the RE remains almost the same during the applied voltage generated by the function generator. The control also helped to minimize any additional drop in the potential caused by the internal resistance of the electrolyte. The cyclic voltammetry (CV) technique was used for electrolyte characterization, and it was carried out in the same electrochemical cell. This technique is usually employed to estimate the potential for optimum plating and to study the deposition process. The CV was performed by applying different scan rate potentials and a nickel surface was chosen in order to study the chemical reactions. The thin nickel film (100 nm thickness) was covered by electroplating tape, except for an area of 0.72 cm^2 and was placed in the WE. The scans were addressed in the cathodic direction (from +0.5 V to −0.15 V vs. SCE). All electrochemical experiments were performed at 25°C, pH 3.67 and used a Potentiast/Galvanostat (VersaStat TM II — Princenton Applied Research) interfaced with a personal computer (PC) and controlled by the Power Suit program.

Figure 8.7a shows the CV curves of the plating bath at different scan rates. As shown in the figure, for most scan rates, when the potential is scanned from +0.5 V to −0.6 V, no considerable current flow is observed. According to the CV, the metal deposition starts at around −0.6 V vs. SCE. As the cathodic potential increases from −0.6, the current also increases and reaches a peak at around −1.0 V. This peak corresponds to the reduction of Ni^{2+} to Ni in the WE. The exponential trend of I at higher absolute values of V might be associated with capacitive charge currents caused by oxygen evolution in the aqueous solution as proposed by Mimani *et al.* [56]. At a scan rate of 156 mV/s, when the potential is reversed, the current follows the same potential dependence up to the irreversibility point of −1.1 V. The irreversibility point increases as the scan rate increases. After the potential reaches back to −0.6 V, negative currents start to flow independently of the scan rate. A new peak appears and its intensity depends on the scan rate potentials. This wide peak corresponds to the oxidation of Ni to Ni^{2+} together with a possible hydrogen evolution reaction in the auxiliary electrode [56]. In general, the CV in the figure is quasi-irreversible, independently of the scan rate. Different scan rates do not change the shape of the cyclic loop but increase the loop area, the peak current and peak potentials separation as shown in the inset in the figure. These characteristics in the CV reveal that the Nerst equation is only approximately satisfied [57]. To obtain the diffusion coefficient (D), we use the Randless-Sevcik equation [58]

$$I_P = 2.69 \times 10^5\, n^{3/2}\, D^{1/2}\, AC\, v^{1/2} \tag{8.5}$$

where the constant C is understood to have units (i.e., 2.69×10^5 C $mol^{-1}v^{1/2}$) and represents the bulk concentration in mol/cm^3, n is the number of electrons transferred, A is the area of the electrode in cm^2, and v is the scan rate in V/s. Therefore, the diffusion coefficient D of the species in cm^2/s is 4×10^{-5} cm^2/s. Based on these results, any applied potential from the interval of −1.2 V to −1.5 V vs. SCE can be selected for a good deposition of Ni. It is obvious that from this interval, the lowest applied potential (−1.2 V) provides the slowest deposition which is preferred for better control of the micro and nano characteristics. Figure 8.3b shows the current and charge transfer variations during electrodeposition at the recommended −1.2 V applied voltage. The current varies exponentially following the Nerst Law [59] and correctly implies that changing the applied potential by several milli-volts drastically modifies the electroplating rate. Typically, an inducing current of 0.2 μA allows an electroplating rate of 1 Å s^{-1} at room temperature [31]. During the experiment, we also observed that a slight increase of the applied potential results in enhanced surface roughness together with a decrease of the density of the deposited metal (data not shown here). From the plot, it is also clear that as time passes, the increase of the surface area allows more charge transfer. The mass transfer obtained by the Faraday law is shown in the inset of the figure, and fitting it with time of the electrodeposition, the following relation is obtained:

$$m = 2.3 \times 10^{-9}\, t^{1.5} \tag{8.6}$$

where m is the deposition mass during the process and t is the time. In this way, it is possible to control the amount of metal deposition on the electrodes by adjusting the time of electrodeposition.

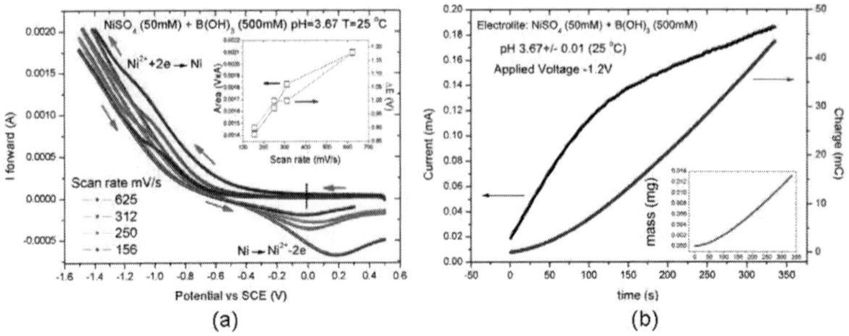

Figure 8.7 (a) Cyclic voltammetry of 50 mM of $NiSO_4$ and 500 mM of H_3BO_3 in aqueous solution at different scan rates. (b) Current and charge variation with time during electrodeposition. Inset (b): deposition mass as function of time. See also Color Insert.

8.3.1.2 Nanogap formation

The initial electrodes were obtained by conventional lithography, four and three metal arrow configurations were obtained by the lift-off technique. In order to focus the plating process on the tips of the electrodes, the samples were spun with a polymethyl methacrylater (PMMA) layer, and a window was opened over the tips. Figure 8.8 shows two samples before electrodeposition. In Fig. 8.8a, four non-uniform nickel electrodes were designed with the horizontal electrodes initially separated by 2.5 μm. The plating was focused only on the tips in the opened rectangular window of dimensions 28 μm × 100 μm. In Fig. 8.8b, three sharper uniform arrows with initial separation of 200 nm were fabricated, and the plating was focused on the electrode tips by opening a circular window of 6 μm diameter. In each case, only the horizontal electrodes were connected to the electrodeposition set-up.

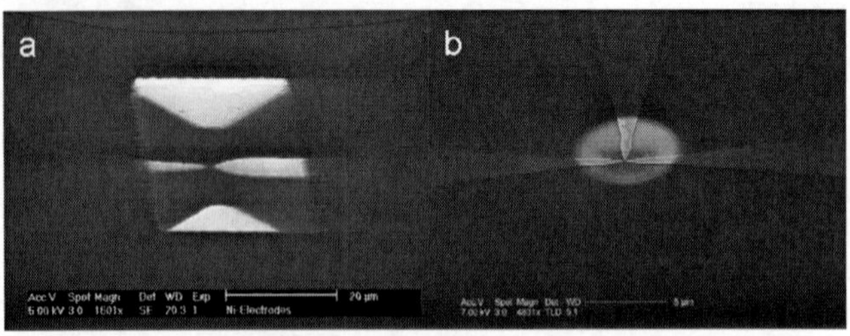

Figure 8.8 Initial nickel electrode configurations: (a) rectangular window and (b) circular window over the electrodes tips. In both cases the plane surfaces are tilted around 53°.

Figure 8.9 presents micrographs of the growth evolution of the type of electrodes configuration with rectangular window described above. As expected, the electrodeposition was focused only on the exposed area of the window. The images reveal typical plated surfaces, in which nickel ions migrate from the electrolyte to the electrodes making them grow. As described above, in the anode of the cell, the nickel looses two electrons and becomes 2+, Ni^{2+}associates with $(SO_2)^{2-}$ in the solution. In the cathode, the negative charge of the surface attracts the positively charged Ni^{2+} from the solution and repels negative charged ions. The process follows Eq. 8.6 and stops when the applied voltage is switched off. Therefore, the electrode growth can be controlled with the applied voltage and stopped at different times to achieve the desired width. In the figure, the initial nickel electrodes have a thickness of 50 nm and were separated by about 2.5 μm. Electrodeposition makes the thickness go up to fifty times and the gap separation decreases with electroplating time until they make contact.

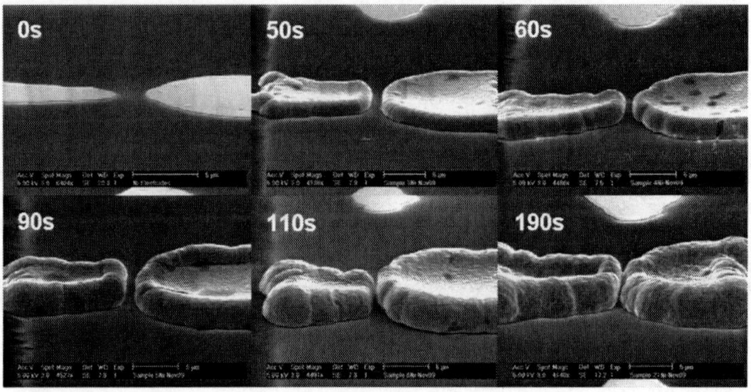

Figure 8.9 Electrode growth during electroplating.

Figure 8.10 shows the nickel electrodes exposed with circular window described above after they are electrodeposited until forming a contact. Note that the vertical electrode which has not been connected to the electrochemical set-up has etched after contact with the electrolyte. Moreover, due to the smaller and more pointed electrode tips than in the previous case, during electroplating, the metal ions of the electrolyte are better guided by the electric field to the sharpest points. Then since the highest gradients of the field are located in the outer corner of each arrow, out-of-plane growth of the electrodes results. As we will see in the next section, contact electrodes obtained by electrodeposition can be easily reopened by the electromigration induced breaking junction (EIBJ) technique, and then out of plane electrodes provide some advantages in that technique: (i) there is no residual interlink of metallic adhesion layer over the SiO_2 preventing the formation of parallel conduction channels on the substrate which can disturb the measurements [60], (ii) there is smaller probability for island formation due to Joule heating [61], and (iii) it avoids the use of molecular adhesion layers beneath the metal [62]. Molecular devices could be obtained by depositing the molecules on the electrodes before breaking and inducing them to migrate into the gap by subsequent heating. In this way, undesired contact between molecules and substrate can be avoided.

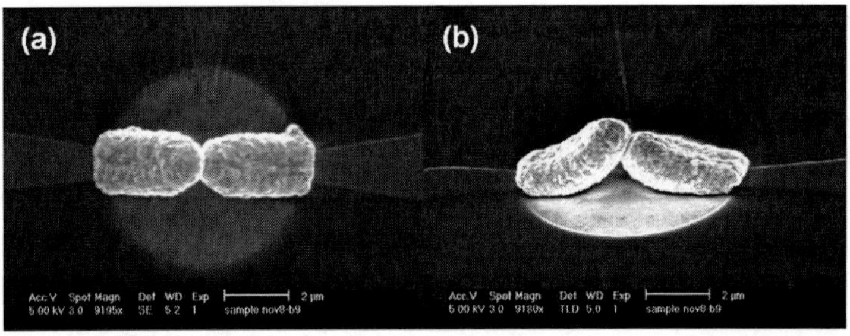

Figure 8.10 Electrochemical plated nickel electrodes until forming contact, (a) top view and (b) lateral view.

At electrodeposition time of 290 s, the formed gap can hardly be seen by the microscope and *I*–*V* measurements need to be performed to detect them as will be discussed in the next section. The electrodeposition procedure described above leads to successful application of Eq. 8.6. It also allows one to control the gap width between the electrodes by setting the time of electrodeposition alone while taking into account other parameter such as electrolyte components, applied voltages, etc. Following this procedure, it is not necessary to average measurements from many samples to find a desired gap width. Figure 8.11 shows gap width variation with time of electrodeposition. Up to 160 s, the gap width decreases exponentially following the trend:

$$d = 2.44e^{-(t/100)} \quad (8.7)$$

where d is the gap width (in microns) and t is the plating time (in seconds). This behavior is expected considering the mass transfer during deposition is not linear with time as discussed above. However, we found that for deposition times higher than 190 s a linear relation dominates the last stages before the electrodes get into contact as shown in the figure inset. In the next section, the current voltage characterization is described.

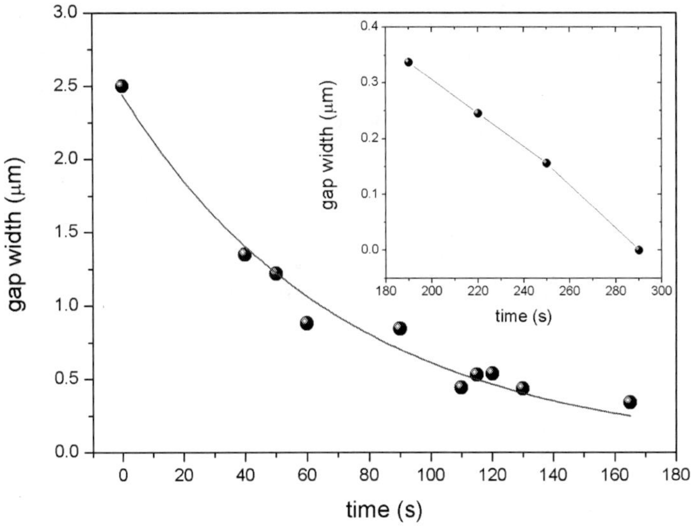

Figure 8.11 Gap width variation with electroplating time.

8.3.2 Characterization

Figure 8.12 shows the Current–voltage response for the Ni samples electrodeposited during 290, 315, and 350 s. We have observed recordable and reproducible *I–V* characteristics in the milliampere and microampere range in repeated scans, which confirm the stability of the samples. In the figure, three different signal behaviors corresponding to different ranges of electrode separation can be identified. In the first case, the dashed line at zero indicates that no current flows which corresponds to a large electrode separation. The second case relates to samples electrodeposited for 290 and 315 s and shows tunneling behavior revealing nanogap formation. In the third case, after 350 s the measurements show Ohmic behavior revealing that the electrodes are in contact. In the tunneling signals, after 290 and 315 s of electrodeposition, the samples allow currents of 0.12 μA (~30 nS conductance) and 8.2 mA (~2 mS conductance), respectively, to flow at 4.0 V. Note that the conductance quantum $G_0 = 2e^2/h$ is ~77.48 μS, and it has been reported that a tunneling conductance in the order of nS is the typical response of gaps with about 1 nm width [6,24,63], and it decreases exponentially as the gap width increases [64]. Therefore, in this work the gap width of the sample electrodeposited for 290 s is around 1 nm, whereas the electrodes electrodeposited for 350 s should have atomic separation. These results obviously confirm that with larger deposition times the gap between the electrodes decreases and the conductance increases, and eventually at 350 s the electrodes make contact. Once the contact is formed, the detected resistance is 157 Ω which implies ~$80G_0$ conductance. This means that around eighty conducting channels have been formed. The contact allows current densities up to around 4.5×10^{-10} A/cm^2 to flow, and at applied voltages higher than 8.0 V, the junction breaks as shown in the stressing curve in the figure inset. Once the junction breaks, a nanogap is again formed as we discuss next.

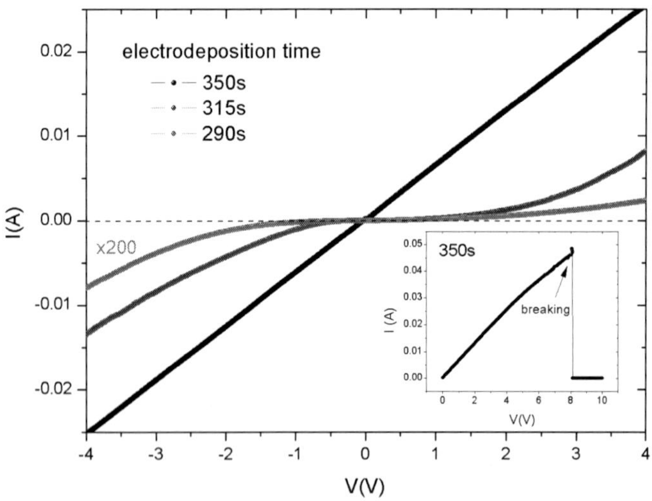

Figure 8.12 Current–voltage responses of the electrodes for 290, 315, and 350 s of electrodeposition. The tunneling behavior reveals that nanogaps are formed at 290 and 315 s, whereas Ohmic behavior indicates that a contact is formed after 350 s of electroplating. Inset: current stressing curve of the sample electrodeposited for 350 s. See also Color Insert.

Nanogaps can also be reopened after the electrodes come into contact by inducing the junction to break by means of the electromigration EIBJ technique. For instance, the resistance of the contact shown in Fig. 8.10 above was 158 Ω. Applying higher voltages than 1.5 V allows electromigration, the contact breaks at a threshold current (I_{th}) of 11 mA. The nanogap is then reopened due to the physical motion of atoms out of the high-current density areas, i.e., the contact [65, 66]. Figure 8.13 shows the *I*–*V* measurement of the reopened gap (blue line in the figure), and tunneling behavior is detected again. Considering the effective frontal area of around 7.8 $\times 10^{-12}$ cm^2, the threshold current density ($J_{th} = I_{th}/A$) for nanogap formation by electromigration is 1.4×10^{-9} A/cm^2. J_{th} values of about 2.9×10^{-8} A/cm^2 have been obtained using gold wires [67], and this implies that J_{th} values of the order of 10^{-8} to 10^{-9} A/cm^2 are required for nanogap formation by EIBJ. Moreover, the breaking signal due to the EIBJ technique confirms that the nanogap has reopened, and hence the *I*–*V* signal shown in the figure results from tunneling

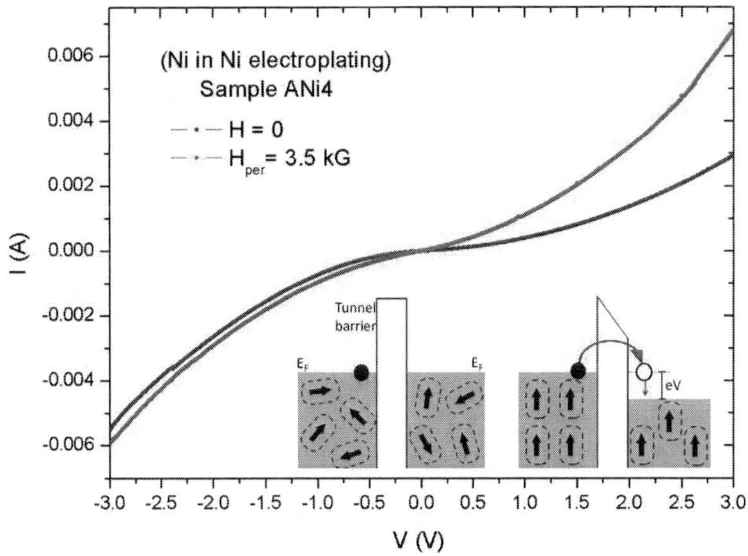

Figure 8.13 *I*–*V* tunneling behavior in a reopened nickel nanogap. Inset: Schematic representation of the tunneling effect in the nickel electrodes. Asymmetry is corrected by the application of an external magnetic field suggesting that charge transfer in the nickel electrodes depends on the orientation of the magnetic moments. See also Color Insert.

through the gap and not through natural oxide layers. To date, there is no exact mechanism known for electron transport through nickel nano-electrodes in nanodevices. Nevertheless, it might be interpreted as follows: when no external magnetic field is applied, the electrodes have randomly oriented magnetic moments and the gap behaves as a tunnel barrier (as represented inset down in the figure). This disorientation should be responsible for the asymmetric tunneling current when the voltage bias is applied. In contrast, when a magnetic field is applied (e.g., 3.5 kG), the magnetic moments align, the resistance is reduced, and this causes the conductance to become symmetric (red line in the figure). Similar results have been obtained by other authors [68,69], and it has also been proposed that under external magnetic fields, the double energy bands of the ferromagnetic electrodes split, increasing the number of sub-bands crossing the Fermi level and, thus, increasing the conductance in the electrodes [28,70].

References

1. Datta S., Tian W., Hong S., Reifenberger R., Henderson J.I., and Kubiak C.P. (1997) Current–voltage characteristics of self-assembled monolayers by scanning tunneling microscopy, *Physical Review Letters,* **79,** pp. 2530–2533.
2. Emberly E.G. and Kirczenow G. (2003) The smaller molecular switch, *Physical Review Letters*, **91,** pp. 188301–188304.
3. Wold D.J. and Frisbie C.D. (2001) Fabrication and characterization of metal-molecule-metal junctions by conducting probe atomic force microscopy, *Journal of the American Chemical Society,* **123,** 5549–5556.
4. Moreland J. and Ekin J.W. (1985) Electron tunneling experiments using Nb-Sn "break" junctions, *Journal of Applied Physics*, **58 (10)**, pp. 3888–3895.
5. Reed M.A., Zhou C., Muller C.J., Burgin T.P., and Tour J.M. (1997) Conductance of a molecular junction, *Science*, **278**, pp. 252–254.
6. Kergueris C., Bourgoin J.P., Palacin S.,. Esteve D, Urbina C., Magoga M., and Joachim C. (1999) Electron transport through a metal-molecule-metal junction, *Physical Review B*, **59**, pp. 12505–12513.
7. Smit R.H.M., Noat Y., Untiedt C., Lang N.D., van Hemert M.C., and van Ruitenbeek J.M. (2002) Measurement of the conductance of a hydrogen molecule, *Nature*, **419**, pp. 906–909.

8. Zhou C., Muller C.J., Deshpande M.R., Sleight J.W., and Reed M.A. (1995) Microfabrication of a mechanically controllable break junction in silicon, *Applied Physics Letters*, **67**, 1160.
9. Krans J.M., van Ruitenbeek J.M., Fisun V.V., Yanson I.K., and de Jongh L.J. (1995) the signature of conductance quantization in metallic point contacts, *Nature*, **375**, pp. 767–769
10. Muller C.J, Krans J.M., Todorv T.N., and Reed M.A. (1996) Quantization effects in the conductance of metallic contacts at room temperature, *Physical Review B*, **53**, pp.1022–1025.
11. Parks J.J., Champagne A.R., Hutchison G.R., Flores-Torres S., Abruña H.D., and Ralph D.C. (2007) Tuning the Kondo effect with a mechanically controllable break junction, *Physical Review Letters*, **99**, 026601.
12. Van Ruitenbeek J.M., Allvarez A., Pineyro I., Grahmann C., Joyez P., Devoret M.H., Esteve E., and Urbina C. (1996) Adjustable nanofabricated atomic size contacts, *Review of Scientific Instruments*, **67**, pp.108–111.
13. Liang W., Shores M.P., Bockrath M., Long Jr., and Park H. (2002) Kondo Resonance in a single-molecule transistor, *Nature*, **417**, pp. 725–729.
14. Bolotin K.I., Kuemmeth F., Pasupathy A.N., and Ralph D.C. (2004) Metal-nanoparticle single-electron transistors fabricated using electromigration, *Applied Physics Letters*, **84**, pp. 3154–3156.
15. Mahadevan M. and Bradley R.M. (1999) Simulations and theory of electromigration-induced slit formation in unpassivated single-crystal metal lines, *Physical Review B*, **59**, 11037.
16. Park H, Park J., Lim A.K., Anderson E.H., Alivisatos A.P., and Mc Euen P.L. (2000) Nanomechanical oscillations in a single-C60 transistor, *Nature*, **407**, pp. 57–60.
17. Taychatanapat T., Bolotin K.I., Kuemmeth F., and Ralph D.C. (2007) Imaging electromigration during the formation of break junctions, *Nano Letters*, **7**, pp. 652–656.
18. Durkan C. (2007) *Current at the Nanoscale: An Introduction to Nanoelectronics*, 1st Ed. (Imperial College Press, UK).
19. Park H., Lim A.K.L., Alivisatos A.P., Park J., and McEuen P.L. (1999) Fabrication of metallic electrodes with nanometer separation by electromigration, *Applied Physics Letters*, **75**, pp 301–303.
20. Khondaker S.I (2004) Fabrication of nanoscale device using individual colloidal gold nanoparticles, *IEE Proceedings – Circuits Devices Systems*, **151**, pp. 457–460.
21. Khondaker S.I. and Zao Z. (2002) Fabrication of nanometer-spaced electrodes using nanoparticles, *Applied Physics Letters*, **81**, pp. 4613–4615.

22. Park J., Pasupathy A.N, Goldsmith J.I, Chang C., Yaish Y., Petta J.R., Rinkoski M., Sethna J.P., Abruna H.D., McEuen P.L., and Ralph D.C. (2002) Coulomb blockade and the Kondo effect in single-atom transistors, *Nature*, **417**, pp. 722–725.
23. Deshmukh M.M., Prieto A.L., Gu Q., and Park H. (2003) Fabrication of asymmetric electrode pairs with nanometer separation made of two distinct metals, *Nano Letters*, **3**, pp. 1383–1385.
24. Li C.Z., He H.X., and Tao J. (2000) Quantized tunneling current in the metallic nanogaps formed by electrodeposition and etching, *Applied Physics Letters*, **77**, pp. 3995–3997.
25. Kervennic Y.V., Van der Zant H.S.J., Morpurgo A.F., Gurevich L., and Kouwenhoven L.P. (2002) nanometer-spaced electrodes with calibrated separation, *Applied Physics Letters*, **80**, pp. 321–323.
26. Yang C.-S., Thiltges J., Doudin B., and Johnson M. (2002) In situ monitoring of quantum conductance in electrodeposited magnetic point contacts , *Journal of Physics: Condensed Matter*, **14**, pp. L765–L771.
27. Kervennic Y.V., Vanmaekelbergh D., Kouwenhoven L.P., and Van der Zant H.S.J. (2003) Planar nanocontacts with atomically controlled separation, *Applied Physics Letters*, **83**, pp. 3782–3784.
28. Sokolov A., Zhang C., Tsymbal E.Y., Redepenning J., and Doudin B. (2007) Quantized magnetoresistance in atomic-size contacts, *Nature Nanotechnology*, **2**, pp. 171–175.
29. Céspedes O., Bari M.A., Dennis C., Versluijs J.J., Jan G., O'Sullivan J., Gregg J.F., and Coey J.M.D. (2002) Fabrication and characterization of Ni nanocontacts, *Journal of Magnetism and Magnetic Materials*, **242–245**, pp. 492–494.
30. Sahoo S., Yang C.-S., and Doudin B. (2006) Fabrication and quantum conductance of electroplated Ni nanocontacts, *Physics Letters A*, **352**, pp. 331–334.
31. Kashimura Y., Nakashima H., Furukawa K., and Torimitsu K. (2003) Fabrication of nano-gap electrodes using electroplating technique, *Thin Solid Films*, **438–439**, pp. 317–321.
32. Morpugo A.F., Marcus C.M., and Robinson D.B. (1999) Controlled fabrication of metallic electrodes with atomic separation, *Applied Physics Letters*, **74**, pp. 2084–2086.
33. Mallory G.O. and Hajdu J.B. (1990) *Electroless plating*, 1st Ed. (Noyes Publications/William Andrews Publishing, USA).

34. Shipway A.N, Katz E., and Willer I. (2000) Nanoparticles arrays on surfaces for electronic, optical, and sensor applications, *European Journal of Chemical Physics and Physical Chemistry*, **1**, pp. 18–52.

35. Li X., Yasutake Y., Kono K., Kanehara M., Teranishi T., and Majima Y. (2009) Au nanoparticles chemisorbed by dithiol molecules inserted in alkanethiol self-assembled monolayers characterized by scanning tunnelling microscopy, *Japanese Journal of Applied Physics*, **48**, 04C180.

36. De Los Santos Valladares L., Bustamante Dominguez A., Llandro J., Suzuki S., Mitrelias T., Bellido Quispe R., Barnes C.H.W., and Majima Y. (2010) Attaching thiolated superconductor grains on gold surfaces for nanoelectronics applications, *Japanese Journal of Applied Physics*, **49**, 093102.

37. Schlesinger M. and Paunovic M. (2010) *Modern Electroplating*, 5th Ed. (John Wiley & Sons. Inc. USA)

38. Okinaka Y. (1970) Electroless gold deposition using borohydride or dimethylamine as reducing agents, *Plating*, **57**, pp. 914–920.

39. Yasutake Y., Kono K., Kanehara M., Teranishi T., Buitelaar M.R., Smith C.G., and Majima Y. (2007) Simultaneous fabrication of nanogap gold electrodes by electroless gold plating using a common medical liquid, *Applied Physics Letters*, **91**, 203107.

40. Azuma Y., Yasutake Y., Kono K., Kanehara M., Teranishi T., and Majima Y. (2010) Single-electron transistor fabricated by two bottom-up processes of electroless Au platting and chemisorption of Au nanoparticle, *Japanese Journal of Applied Physics*, **49**, 090206.

41. Shu C., Li C.Z., He H.X., Bogozi A., Bunch J.S., and Tao N.J. (2000) Fractional conductance quantization in metallic nanoconstrictions under electrochemical potential control, *Physical Review Letters*, **84**, pp. 5196–5199.

42. Garcia N., Wang H., Cheng H., and Nikolic N.D. (2003) Ballistic magnetoresistance versus magnetostriction effects in electrodeposited nanocontacts at room temperature, *IEEE Transactions on Magnetics*, **39**, pp. 2776–2781.

43. Hua S.Z. and Chopra H.D (2003) 100,000% ballistic magnetoresistance in stable Ni nanocontacts at room temperature, *Physical Review B*, **67**, 060401.

44. Mallett J.J., Svedberg E.B., Ettedgui H., Moffat T.P., and Egelhoff W.F. (2004) Absence of ballistic magnetoresistance in Ni contacts controlled by an electrochemical feedback system, *Physical Review B*, **70**, 172406.

45. Svedberg E.B., Mallett J.J., Ettedgui H., Gan L., Chen P.J., Shapiro A.J., Moffat T.P., and Egelhoff W.F. (2004) Resistance changes similar to ballistic magnetoresistance in electrodeposited nanocontacts, *Applied Physics Letters*, **84**, pp. 236–238.
46. Ozatay O., Chalsani P., Emley N.C., Krivorotov I.N., and Burhman R.A (2004) Magnetoresistance and magnetostriction effects in ballistic ferromagnetic constrictions, *Journal of Applied Physics*, **95**, Part 2, pp. 7315–7317.
47. Elhoussine F., Mátéfi-Tempfli S., Encinas A., and Piraux L. (2002) Conductance quantization in magnetic nanowires electrodeposited in nanopores, *Applied Physics Letters*, **81**, pp. 1681–1683.
48. Gabureac M., Viret M., Ott F., and Fermon C. (2004) Magnetoresistance in nanocontacts induced by magnetostrictive effects, *Physical Review B*, **69**, 100401.
49. Egelhoff W.F., Gan L., Ettedgui H., Kadmon Y., Powell C.J., Chen P.J., Shapiro A.J., McMichael R.D., Mallett J.J., Moffat T.P., Stiles M.D., and Svedgerg E.B. (2004) Artifacts in ballistic magnetoresistance measurements, *Journal of Applied Physics*, **95**, pp. 7554–7559.
50. Wu Y., Chang D., Kim D., and Kwon S.-C. (2003) Influence of boric acid on the electrodepositing process and structures of Ni-W alloy coating, *Surface and Coatings Technology*, **173**, pp. 259–264.
51. Horkans J. (1979) On the role of buffers and anions in NiFe electrodeposition, *Journal of the Electrochemical Society*, **126**, pp. 1861–1867.
52. Popov B.N., Tin K.-M., and White R.E (1993) Galvanic pulse and pulse reverse plating of nickel-iron alloys from electrolytes containing organic compounds on rotating disc electrode, *Journal of the Electrochemical Society*, **140**, pp. 1321–1330.
53. Grand W.C. and Talbot J.B. (1993) Electrodeposition of thin films of nickel-iron, *Journal of the Electrochemical Society*, **140**, pp. 675–681.
54. Karwas C. and Hepel T. (1989) Morphology and composition of electrodeposited cobalt-zinc alloys and the influence of boric acid, *Journal of the Electrochemical Society*, **136**, pp. 1672–1678.
55. Tilak B.V., Gendron A.S., and Mosoiu M.A. (1977) Borate equilibria in nickel refining electrolytes, *Journal of Applied Electrochemistry*, 7, pp. 495–500.
56. Mimani T., Mayanna S.M., and Munichandraiah N. (1993) Influence of additives on the electrodeposition of nickel from a Watts bath: A cyclic voltammetric study, *Journal of Applied Electrochemistry*, **23**, pp. 339–345.

57. Heinze J. (1984) Cyclic voltammetry — "electrochemical spectroscopy". New analytical methods (25), *Angewandte Chemie International Edition in English*, **23**, pp. 831–847.

58. Hamann C H, Hamnett A., and Vielstich W. (2007) *Electrochemistry*, 2nd Ed. (Willey-VCH, Weinheim, Germany).

59. Bard A.J. and Faulkner L.R. (2001) *Electrochemical Methods, Fundamentals and Applications*, 2nd Ed. (John Wiley & Sons Inc., New York, USA).

60. Gabrielson T.B. (1993) Mechanical-thermal noise in micromachined acoustic and vibration sensors, *IEEE Transactions on Electron Devices*, **40**, pp. 903–909.

61. Trouwborts M.L., van der Molen S.J., and van Wees J. (2006) The role of Joule heating in the formation of nanogaps by electromigration, *Journal of Applied Physics*, **99**, 114316.

62. Mahapatro A.K., Ghosh S., and Janes D.B (2006) Nanometer scale electrode separation (nanogap) using electromigration at room temperature, *IEEE Transactions on Nanotechnology*, **5**, pp. 232–236.

63. Reichert J., Ochs R., Beckmann D., Weber H.B., Mayor M., and Löhneysen H.V. (2002) Driving current through single organic molecules, *Physical Review Letters*, **88**, 176804.

64. Simmons J.G. (1963) Generalized formula for the electric tunnel effect between similar electrodes separated by a thin insulating film. *Journal of Applied Physics*, 34, pp. 1793–1803.

65. Tu K.N (2003) Recent advances on electromigration in very-large-scale-integration of interconnects, *Journal of Applied Physics*, **94**, pp. 5451–5473.

66. Ho P.S. and Kwok T. (1989) Electromigration in metals, *Reports on Progress in Physics*, **52**, pp. 301–348.

67. Hadeed F.O. and Durkan C. (2007) Controlled fabrication of 1–2 nm nanogaps by electromigration in gold and gold-palladium nanowires, *Applied Physics Letters*, **91**, 123120.

68. Céspedes O., Rocha A.R., Lioret S., Viret M., Dennis C., Gregg J.F., van Dijken S., Sanvito S., Coey J.M.D., (2004) *I–V* asymmetry and magnetoresistance in nickel nanoconstrictions, *Journal of Magnetism and Magnetic Materials*, **272–276**, pp. 1571–1572.

69. De Los Santos Valladares L., Leon Felix L., Bustamante Dominguez A., Mitrelias T., Sfigakis F., Khondaker S.I., Barnes C.H.W., and Majima Y. (2010) Controlled electroplating and electromigration in nickel electrodes for nanogap formation, *Nanotechnology*, **21**, 445304.

70. Johnson M. (2007) Magnetoelectronics: New twist for ballistic transport, *Nature Nanotechnology*, **2**, pp. 143–144.

Chapter 9

Nanometer-Scale Processing by Tribological Action and Its Potential Applications

Shojiro Miyake[a] **and Mei Wang**[b,*]
[a] *Department of Innovative Systems Engineering, Nippon Institute of Technology, Miyashiro-machi, Saitama, 345-8501, Japan*
[b] *Department of Research and Development, OSG Corporation, 1-15 Honnogahara, Toyokawa, Aichi 442-8544, Japan*
*mwang@osg.co.jp, saburo@leo.nit.ac.jp

9.1 Introduction

It is thought that various nanoelectronic devices and nanomachines can potentially be manufactured by manipulating atoms and molecules [1]. Microfabrication technology is the key to the development of these nanotechnologies and is the first challenge to be overcome. For conventional microfabrication, lithography is the most commonly used semiconductor manufacturing process technology, which is usually combined with deposition and dry and wet etching processes. More recently, with the development of nanotechnology, new nanoprocessing methods such as nanoprinting and molecular self-organization have been developed. However, these

Vistas in Nanofabrication
Edited by Faiz Rahman

ISBN 978-981-4364-56-0 (Hardcover), 978-981-4364-57-7 (eBook)
www.panstanford.com

nanoprocessing methods are limited by the shape and dimensions of workpiece materials.

On the other hand, using the tribological behavior of tool geometry and a workpiece, processing to obtain high-precision three-dimensional microshapes can potentially be performed by a mechanical processing method transcribing a tool locus [2]. For example, a 1 μm-wide knife edge can be formed by fine abrasive grinding [3]. Using microtribological action, the degree of freedom of processed objects such as the material, shape, and size will increase if shape processing in micro- and nanometer-scale areas can be realized. It is, therefore, realized that the application of microfabrication can be significantly extended through such novel processes. To illustrate the technological innovation in fine-shape processes, examples where nanofabrication has been performed using an atomic force microscope (AFM) based on the tribological action between the tip and a processed workpiece and where three-dimensional fine-shape processes at the nanometer scale have been realized will be described in this chapter.

9.2 Nanoprocessing Methods and Tribology

9.2.1 AFM Nanoprocessing

The ultimate aim of microfabrication is to realize a process to remove atoms and molecules from predetermined locations or attach them to predetermined locations. In other words, atomic manipulation technology rearranges atoms by moving them from one place to another. Meanwhile, scanning tunneling microscopy (STM) is widely used as a means of observing surface at the atomic scale [4]. Scanning probe microscopy (SPM) has been developed as a method of measuring and observing various types of information. SPM technology is also very useful for evaluating the properties of nanoscale surfaces. For example, atomic force microscopy (AFM) is used for measuring and evaluating mechanical characteristics at the nanoscale [5].

In SPM studies of microfabrication, atomic manipulation such as the fabrication of atomic characters has been achieved [6]. Microscopic manufacturing methods comprising discharge and evaporation have been proposed and applied for the manipulation of atoms. SPM is a promising method for the nanofabrication of functional nanometer-scale engineering materials and devices. SPM nanoscale processing technology has been focused on as a means of achieving nanolithography for the fabrication of nanomachines and nanodevices. As a result, by using AFM to measure the microscopic atomic forces based on probe technology, nanoscale mechanical processing can be realized [7]. Determining the limits to the use of microfabrication in machining is the most important issue. Wear in units of atomic layers has been observed as a tribological phenomenon [8]. If the wear can be controlled, processing at the atomic level by machining will become possible. Machining at the level of individual layers for layered crystal materials has been performed [9].

Figure 9.1 shows the AFM nanoprocess method [10]. AFM can detect interactions between atoms. By increasing the atomic force applied to a hard diamond tip in order to induce its motion relative to a workpiece, nanomachining can be achieved. Furthermore, the scanning path and the atomic force of a superhard-film-coated tip can be controlled by adjusting the voltage of the AFM piezoelectric device by a computer to make shape processing possible. By optimizing the atomic force on the workpiece, micromachining becomes possible at this point as the processing unit is controlled at the atomic level. In these micromachining processes, diamond-like carbon (DLC) and carbon nanotubes (CNTs) are considered suitable tool materials [11, 12].

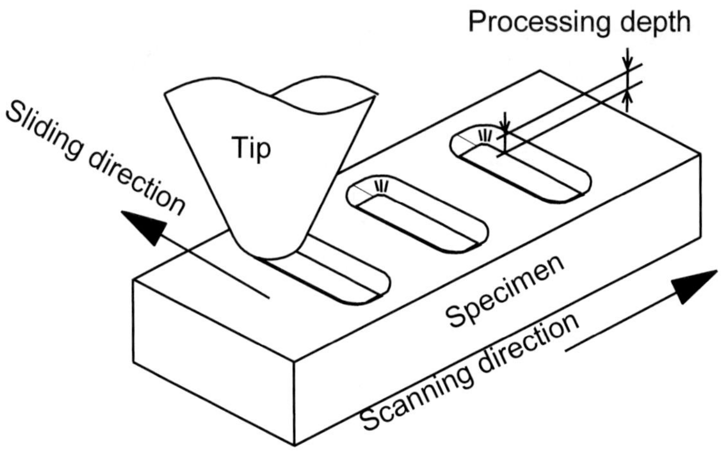

Figure 9.1 Nanofabrication using atomic force microscopy.

9.2.2 Nanotribology and Nanofabrication

In the areas of micro- and nanotechnology, nanotribology is important for achieving smooth motion. From the microscopic viewpoint, friction is the interaction between opposing surfaces of each atom, and wear is bonding and destruction of bonds between surface atoms. However, these particle properties of matter have been ignored in traditional tribology, which assumes a homogeneous continuum model. In response, nanotribology is a novel concept dealing not with continuous wear but rather with discontinuous wear at the atomic level [13]. In other words, it is a manifestation of the particle properties of individual atoms [13]. Furthermore, if the wear can be controlled, this can be considered to be nanoprocessing by utilizing the phenomenon of nanotribology. The degree of freedom of processed objects such as their material, shape, and size will increase if shape processing in micro- and nanometer areas can be realized. This would be a major contribution to the development of nanotechnology [14].

In nanotribology, the wear properties of various materials have been investigated. For instance, wear tests have been performed on ion-implanted silicon wafers, ion-implanted diamond films, and nitrogen-containing DLC films [15]. In these tests, a high-precision square groove with an area of $1 \times 1\ \mu m^2$ and a depth of several nanometers, a line, and a space were formed.

To realize high-precision machining, factors such as smoothness, constant intensity, and inert and nonadhesive properties of the materials involved are essential. In our previous studies [16,17], where a DLC film was used as a workpiece, a very precise shape can be formed. Figure 9.2 shows very fine line-scratched grooves and a rectangular groove on a DLC film [16,17]. The line-scratched grooves were formed by tip scanning owing to the friction force. The fine rectangular groove shown in Fig. 9.2 was processed at a load of 20 μN, a feed of 1 μm and a sliding cycle of 100 and was formed by 1 to 36 scanning passes. In this case, about one atomic layer was removed by one-pass scanning. As a result of machining performed by AFM, a processed groove with a very smooth bottom surface was obtained with an R_{max} of below 1 nm. In addition, the microwear of the DLC film was evaluated from the difference in wear depth to be at a scale of about an atomic radius [18]. This demonstrates that microprocessing is possible at the atomic level.

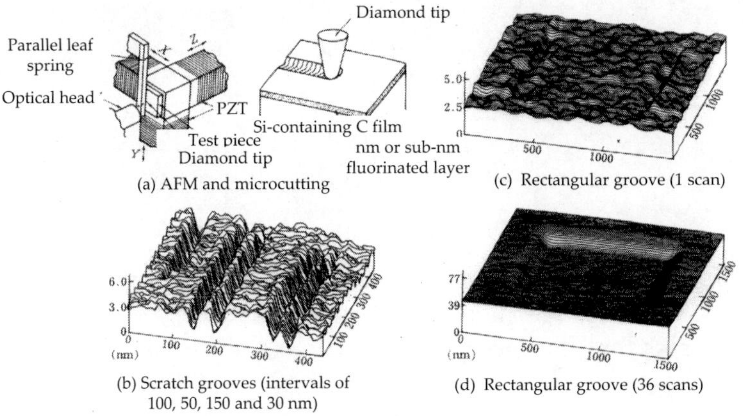

Figure 9.2 Low-energy surface and grooves formed by SPM microcutting.

9.3 Tribomechanical Processing and Application of Etching Mask for Silicon

As a method for nanodevice fabrication, SPM-discharged local oxidation has attracted attention as a microfabrication process for electronic devices [19]. However, the performance of such processing is greatly affected by the surrounding atmosphere and the condition of the surface adsorption layer, and, therefore, the stability of processing has become an issue. Meanwhile, not only wear removal but also protuberances are generated by sliding a diamond tip on a silicon wafer under certain processing conditions. By applying the protuberance phenomenon, protuberance processing characteristics and their mechanism for silicon processing have been investigated [20]. Since the processed area is oxidized, the oxidization layer formed can act as a mask for an etching solution of potassium hydroxide (KOH) during silicon processing [21].

Protuberance and groove processing were performed on a silicon sample in air using a diamond tip of 100 nm radius. The processed area was 1×1 μm^2, and the loads used for processing ranged from 10 to 90 μN. Nine areas were processed gradually increasing by steps of 10 μN. As shown in Fig. 9.3, for loads between 10 and 30 μN, the heights of the formed protuberances were less than 1 nm. For loads of 40, 50, and 60 μN, the heights increased to 1–2 nm. The protuberance heights increased with increase in applied load. This phenomenon can be explained by the model of protuberance processing shown in Fig. 9.4a, where a protuberance is formed by sliding a diamond tip with a radius of 100 or 200 nm. As a consequence of the mechanochemical action caused by the tip sliding in the processed area, the destruction of local bonds in the processed area occurs owing to the action of maximum shear stress or maximum tensile stress. As a result, it is thought that as the reaction of silicon with water, which is present in the atmosphere or adsorbed on the surface of the processed area, forms silicon oxide or silicon hydroxide as chemical reactants, the processed area starts to protuberate. Furthermore, as the maximum tensile stress and the maximum shear stress increases with increasing the processing load, the height of the protuberances also increases.

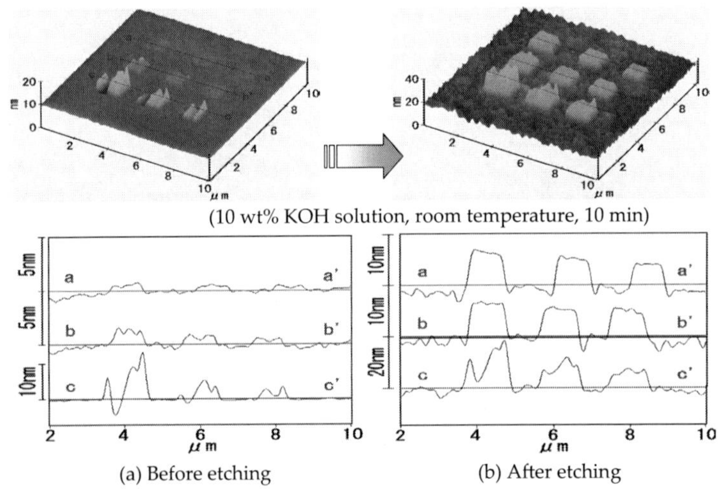

Figure 9.3 Application of etching mask in Si nanoprocessing through mechanochemical action.

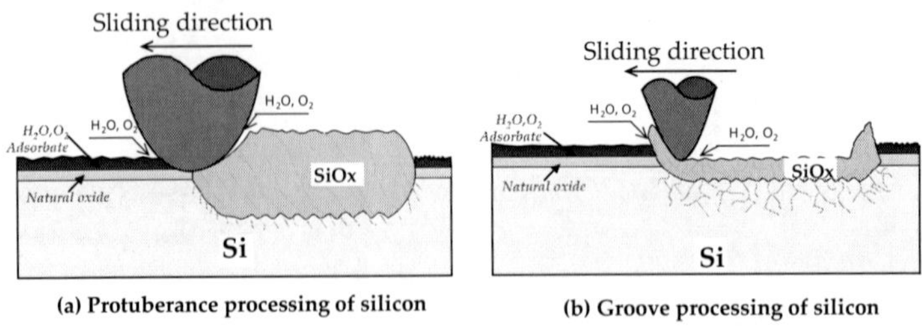

Figure 9.4 Mechanistic models of protuberance and groove processing with a diamond tip.

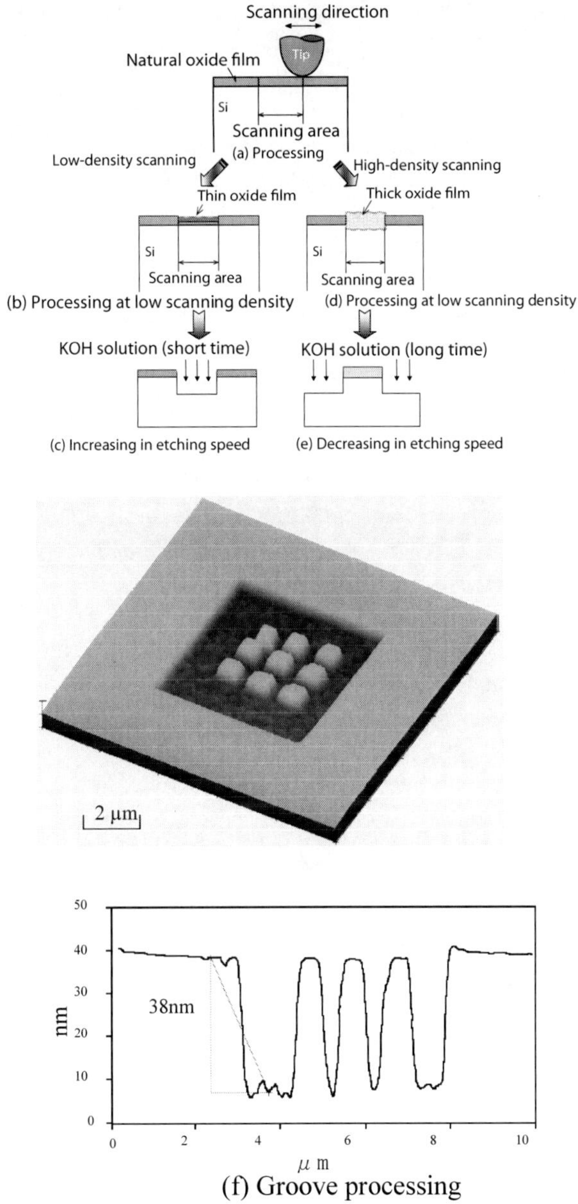

Figure 9.5 Acceleration and deceleration of etching by tribological processing (5 wt% KOH solution, room temperature, and 20 min etching.

For a load of 70 μN, some plastic deformation occurred, which was then removed. The maximum height of the processed area was about 4.5 nm [22]. Since the von Mises stress increases with load, the plastically deformed part generated in front of the indenter is removed instead of the whole protuberance in processed area [22]. In this case, the oxide layer is formed by the same mechanochemical action based on the reaction with water and oxygen as that in protuberance processing. The oxidized portion of the processed area acts as a mask in the KOH solution etching of silicon. Figure 9.5 shows an example of a small portion of a large sample formed by oxidative deterioration under certain conditions being used as a mask in an etching process with KOH solution [22]. In addition, acceleration and deceleration of the rate of etching are possible by changing the scanning density of tip sliding. In the example mentioned above, nine areas of 1×1 μm^2 were scanned with a high scanning density of tip sliding, and oxidative layers were formed due to mechanochemical action. Following this, the processed surface was scanned over an area of 5×5 μm^2 at a low scanning density of tip sliding and then processed by etching with KOH solution as shown in Fig. 9.5 [23]. In the case of a short etching time with KOH solution, the area processed at a high scanning density and the unprocessed area covered with a natural oxidative layer acted as masks and were not etched. In this way, mechanochemical action enables various types of nanofabrication processes to be realized. Nanoscale local oxidation based on mechanochemical action can be expected to be applied to nanodevice processes and nanolithography technology in the future.

9.4 Nanofabrication and its Application based on the Structure of the Workpiece Material

9.4.1 Nanoprocessing of Layered Crystal Materials at the Layer Unit

Utilizing the mechanical anisotropy of crystal materials, if layered crystal materials having periodically weak bonds are used as processed workpieces, processing at the level of single layers can be realized [9]. Mica (Muscovite: Muscovite mica), graphite, molybdenum disulfide (MoS_2), and boron nitride have a layered structure, and there is little interaction between the cleavage planes existing in the basal plane of these materials. Moreover, it is easy to image the atoms on the basal plane, where the processed shape at the atomic order can also be observed.

By applying AFM, the nanomachining characteristics of the layered crystal materials MoS_2, highly-oriented pyrolytic graphite (HOPG) and mica have been investigated using a superhard film tip. MoS_2 has a hexagonal structure [24], and there are van der Waals forces between adjacent S layers. The interval between the cleavage planes is 0.616 nm. HOPG has a structure similar to a single crystal, and the interval between graphite layers is 0.34 nm. Mica has a structure in which O, Al, OH, and K are stacked with a K layer sandwiched between SiO_4 cleavage planes. The layer period is 0.7 nm for the SiO_4 layer and 1 nm for the K layer, and there are van der Waals forces between K and SiO_4 layers.

A DLC-coated Si tip with a tip radius of less than 50 nm was used as a tool for the nanoprocessing of layered crystalline materials. Upon loading a certain atomic force on the DLC-coated Si tip, nanoprocessing is performed by controlling the tip scanning path with a computer. Figure 9.6a shows an example of processing MoS_2 to obtain latticed grooves. The depth of the grooves is about 0.6 nm. This approximately corresponds to the cleavage plane interval of MoS_2 of 0.616 nm, i.e., single-layer processing can be realized. Compared with MoS_2, as shown in Fig. 9.6b, mica can be comparatively easily processed at the layer-unit scale. In contrast, HOPG is difficult to process using the DLC nanofilm tip because the bond strength of the basal plane of HOPG is large.

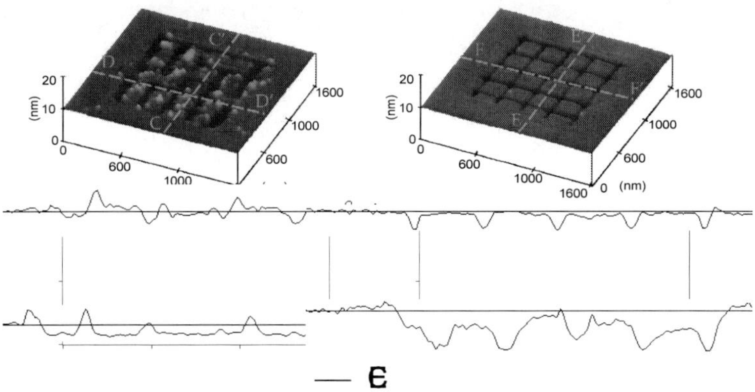

(a) MoS_2 (c) Mica

Figure 9.6 Processing of each layer of layered crystalline materials.

The cutting work of graphite is difficult because the bond strength of its basal plane is extremely high. Conversely, on the basis of the fact that the basal plane of graphite is supported by van der Waals forces, processing involving folding has been investigated. Figure 9.7a shows the surface of HOPG. A step can be seen at the center of the surface. While scanning a diamond tip on the surface under a load, although the surface was not continuously processed, the area of the step was easily transformed and rolled up as shown in Fig. 9.7b. From the cross-sectional profile of the step, the height of the original step was 3.1 nm, the value of which approximately corresponds to nine layers of graphene. These nine graphene layers were formed on the HOPG. The height of a bent part in the step area is about 18 layers [24]. Instead of being destroyed and removed, the bent part was not deformed during tip scanning for shape observation but remained as it was.

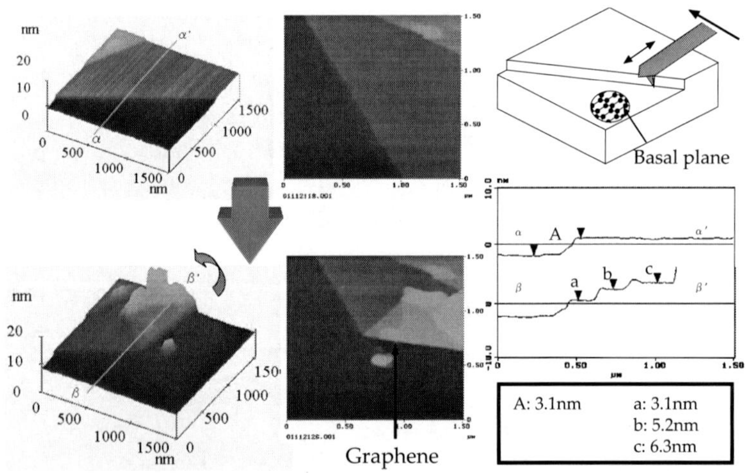

Figure 9.7 Graphene formation from graphite step by diamond tip sliding.

9.4.2 Nanoprocessing of Nanoperiod Multilayer Films

As shown in Fig. 9.8a–c, a nanoperiod multilayer film consists of layers of nanometer order. As a film is layered with nanometer period, it is possible to improve its hardness and microwear characteristics. As shown in Fig. 9.8d, for a 2 nm-period (CN/BN)n multilayer film, the 1 nm-thick boundary of the multilayer exactly corresponds to the

1 nm thickness of each of CN and BN layers. This is the reason for the wear and the mechanism for the suppression of crack propagation [25]. Since these multilayer films are similar to layered crystalline materials, these phenomena show that removal processing can be easily performed at the single-layer scale. Taking advantage of this property, it is possible to perform high-precision microprocessing for each layer of the multilayer film in the same way as that used for the layered crystal material described above.

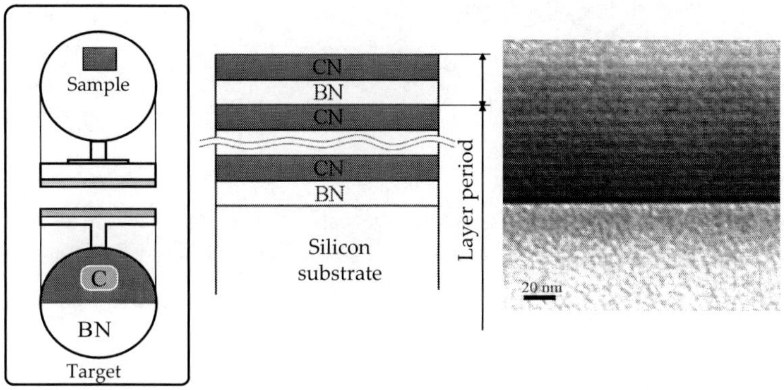

(a) Coating equipment (b) Film structure model (c) Tem image of 8-nm-period nano-layered film

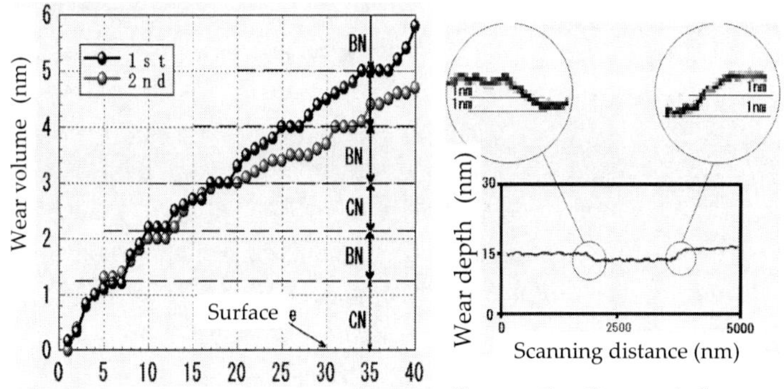

Cross section (2nm-period)

(d) Dependence of the number of friction cycles on microwear

Figure 9.8 Nano-layered film and nano-processing.

Carbon and boron nitride 2 nm-period multilayer films were prepared by reactive sputtering and processed at the nanoscale by AFM as shown in Fig. 9.9 [26]. The nanoprocessing of a groove in each layer was performed at a 1 nm lamination scale corresponding to the thickness of each film. Three adjacent squares of $2 \times 0.7\ \mu m^2$ were scanned at loads of 10, 15, and 20 μN. As a result, each groove was processed to a depth corresponding to about 1 nm lamination thickness, and a three-staged groove was formed.

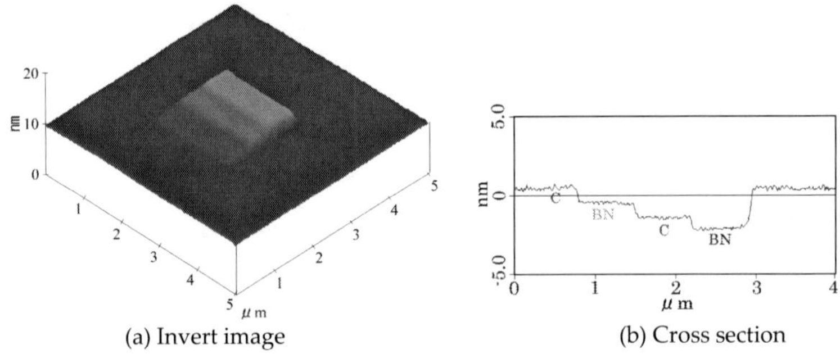

Figure 9.9 Three-staged grooves processed using the structure of a nanoperiod layered film (2 nm-period (CN/BN)n film, loads of 10, 15, and 20 μN).

9.4.3 Application of AFM to Ultrahigh-Density Memory

It has been claimed that as a mainstream of current memory technology, magnetic and optical storage is reaching its physical limits. For this reason, as a possible method of realizing terabit memory (Tb/inch^2), SPM memory has been attracting attention. AFM has been studied as the most viable system for the early realization of high-density memory [27–29]. The tribo-memory method of forming asperities on a recording medium's surface by mechanical action has been used from ancient times. An advantage of this method is its high stability and reliability.

Figure 9.10 shows examples of analog information recording. The initial letters NIT of Nippon Institute of Technology were formed in an area of 100×100 nm^2 with a depth of 1 nm as shown in Fig. 9.10a, and Mr. Tiger's face was depicted in an area of 200×200 nm^2 as shown in Fig. 9.10b [28]. By applying this method, 10 billion characters and 2.5 billion faces can be recorded in an area of 1×1 cm^2. The depth of each processed groove is approximately 1 nm. Corresponding to the period of the cleavage plane in the atomic structure of mica, 1 nm-deep processing was performed.

(a) Nano-characters (b) Nano Mr. Tiger

Figure 9.10 Processing of nanocharacters and nano Mr. Tiger demonstrating analog information recording.

Moreover, as a feasible method of realizing high-density recording tribo-memory, using carbon and carbon nitride nanoperiod multilayer films, the recording method of removing each layer by mechanical action to form grooves with a certain depth was investigated [26,27]. In the tribo-memory, since the limit of miniaturization is determined by the unit and the stability of plastic deformation, the recording density is limited. On the other hand, by utilizing the nano-structure of processed materials, 1 nm-deep processing can possibly be realized, indicating the possibility of atomic-order three-dimensional micromachining.

9.4.4 Application of Nano and Micromachines

In the context of nanotechnology, along with proposed methods for the creation of machines by assembling atoms and molecules, the investigation of exploration engineering for imaging future theoretical nanomachines has been carried out. A molecular bearing similar to a flagella motor has been studied. According to a simulation, a molecular bearing is supported by weak bonds between molecules, i.e., by van der Waals forces, and it is lubricated by the intermolecular force. However, the performance of the bearing supported by van der Waals forces has not experimentally been confirmed. A nanoslider supported by van der Waals forces has been formed owing to the advantageous layered crystal structure of mica, and the result of a study on its behavior and performance is introduced here. Figure 9.11a shows a model of a nanoslider [30], where a weak bond between SiO_4 and K, i.e., van der Waals forces, is at work. The nanoslider is a microscopic specimen supported by van der Waals forces. Moreover, the tip used for processing simultaneously acts as a driver for the nanoslider.

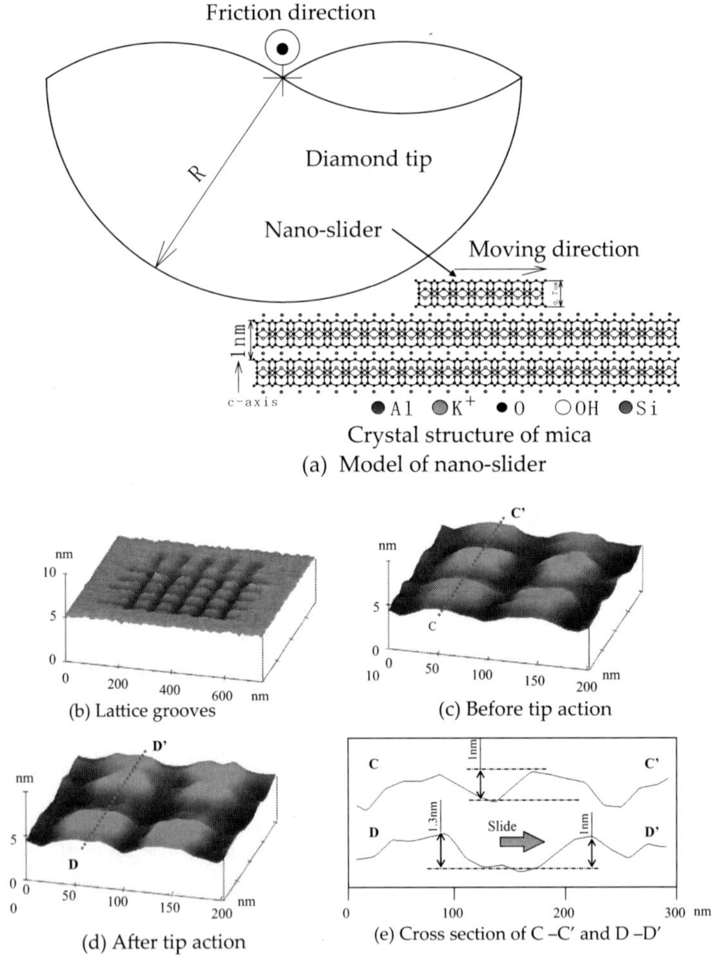

Figure 9.11 Motion of nanoslider supported by van der Waals forces. See also Color Insert.

Figure 9.11b shows 1 nm-deep lattice grooves arranged with a line interval of 80 nm. As shown in Fig. 9.11c, the 50 nm-long nanosliders that were formed were slid by tip sliding twice at a load of 1500 nN, and then the processed area was measured by tip scanning at a load of less than 5 nN as shown in Fig. 9.11d. It can be seen that the two nanosliders formed in the center of the processed area were moved upward along the line owing to friction. Figure 9.11e shows cross sections along lines C–C′ and D–D′. The top layer of the nanoslider, which corresponds to the 1 nm-high upheaval area, slips and is displaced in the direction of D′ by the action of friction. In other words, it can be understood that the SiO_4 top layer merged with the SiO_4 below by the movement of van der Waals bond [30].

9.4.5 Application of Standard of Length Scale

As position resolution at the nanometer scale is necessary in the deposition mechanism of the tool used for forming patterns in semiconductor integrated circuits, SPM high-resolution position sensing can be applied for this purpose. This can be done by utilizing layered crystal materials such as mica and their anisotropic character. If these layered crystal materials with a periodically weak bond are used as a workpiece, processing in the depth direction at the level of single layers can be achieved. Otherwise, the observation of atomic images is possible at these basal planes, and a three-dimensional, high-precision, criteria-scale can be configured by obtaining the crystal lattice spacing. As shown in Fig. 9.12, if a line and space pattern is formed in a mica crystal layer to a depth of 1 nm [31], the scale of the crystal lattice spacing and a two-stepped standard scale of sub-nanometer and sub-micrometer can be realized.

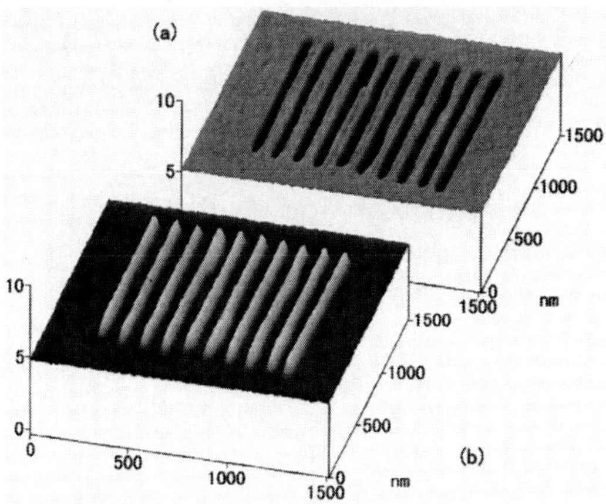

Figure 9.12 Processing of mica surface by cBN-formed tip.

9.5 Conclusions

In this chapter, the authors introduced their research results focusing on nanoprocessing and nanofabrication using the nanotribological action of AFM. We hope that these descriptions will contribute to the further development of this area.

References

1. Drexler, K.E. (1992). *Nanosystems*, John Wiley & Sons Inc.
2. *J. Jpn. Soc. Precision Eng. Ed.* (1993). Nanometer-scale processing technology, Nikkankogyo Newspaper Publishing Company, ISBN 4526033758, 9784526033759, p. 332 (in Japanese).
3. Miyake, S. (1983). Performances of the precision grinding wheels containing solid lubricants, *J. Jpn. Soc. Lubrication Eng.*, 28, 6, p. 419.
4. Binning, G., and Rohrer, H. (1983). Scanning tunneling microscopy, *Surf. Sci.*, 126, p. 236.
5. Miyake, S. (2008). Nanomechanical properties evaluation of surface, *J. Jpn. Soc.* Precision Eng., 73, 8, p. 859 (in Japanese).
6. Eigler, D. M., and Schweizer, E. K. (1990). Positioning single atoms with scanning tunneling microscope, *Nature*, 334, p. 524.
7. Miyake, S. (1993). Nanometer scale thin solid film deposition technology, *J. Jpn. Soc. Precision Eng.*, 59, 4, p. 563. (in Japanese)
8. Miyake, S. (1994). Atomic-scale wear properties of muscovite mica evaluated by scanning probe microscopy, *Appl. Phys. Lett.*, 65, 8, p. 980.
9. Miyake, S. (1995). 1 nm deep mechanical processing of muscovite mica by atomic force microscopy, *Appl. Phys. Lett.*, 67, 13, p. 2925.
10. Miyake, S., Ishi, M., Otake, T., and Tsushima, N. (1997). Nanometer-scale mechanical processing of muscovite mica by atomic force microscopy, *J. Jpn. Soc. Precision Eng.*, 63, 3, p. 426 (in Japanese).
11. Miyake, S. (2004). Nanometer scale processing by tribological action and its potential applications, *J. Vac. Soc. Jpn.*, 47, 12, p. 811 (in Japanese).
12. Miyake S. (2002), Technical Description: Nano-fabrication by AFM using new diamond, *Japan New Diamond Forum*, 67, 18, 4, p. 13 (in Japanese).

13. Kaneko, R., Oguti, S., Miyamoto, Y. Andoh, T., and Miyake, S. (1990). Microtribology for magnetic recording, *STLE Special Publ.*, sp-27, p. 31.
14. Miyake, S. (2005). Tribology of carbonitride and boron nitride films and its application to nanoprocessing, *Thin Solid Films*, 493, 1–2, p. 160.
15. Miyake, S., Watanabe, S. Miyazawa, H., Murakawa, M., Kaneko, R., and Miyamoto, T. (1994). Micro scratch hardness increasing of ion plated carbon film by nitrogen inclusion evaluated by atomic force microscopy, *Appl. Phys. Lett.*, 65, 9, p. 3206.
16. Miyake, S., Miyamoto, T., and Kaneko, R. (1993). Microtribological improvement of carbon film by silicon inclusion and fluorination, *Wear*, 168, p. 155.
17. Miyake, S., Kaneko, R. Kikuya, Y., and Sugimoto, I. (1991). Micro-tribological studies on fluorinated carbon film. *Trans. ASME J. Tribol.*, 113, p. 384.
18. Miyake, S. Kurosaka, W., and Oshimoto, K. (2009). Nanometer-scale mechanical properties of extremely thin diamondlike carbon films, *Tribol. Mater, Surf. Interfaces*, 3, 4, p. 158.
19. Dagata, J. A., Schneir, J., Harary, H. H., Evans, C. J., Postek, M. T., and Bennett, J. (2001). Modification of hydrogen-passivated silicon by a scanning tunneling microscope operating in air, *Appl. Phys. Lett.*, 56, 5, p. 2001.
20. Miyake, S., and Kim, J. (1999). Micro protuberance processing of silicon by diamond tip scanning, *J. Jpn. Soc. Precision Eng.*, 65, 12. p. 1725.
21. Miyake, S., and Kim, J. (2001). Fabrication of silicon utilizing mechanochemical local oxidation by diamond tip sliding, *Jpn. J. Appl. Phys.*, 40, 11B, p. L1247.
22. Miyake, S., and Kim, J. (2005). Nanofabrication of silicon by mechano-chemical processing and its application to etching mask for potassium hydroxide solution, *Nanotechnology*, 16, p. 149.
23. Miyake, S., and Kim, J. (2003). Nanoprocessing of carbon and boron nitride nanoperiod multilayer films, *Jpn. J. Appl. Phys.*, 42, 3B, p. L322.
24. Miyake, S., and Matsuzaki, K. (2002). Mechanical nanoprocessing of layered crystal structure materials by atomic force microscopy, *Jpn. J. Appl. Phys.*, 41, 9A, p. 5706.
25. Miyake, S. (2003). Improvement of mechanical properties of nanometer period multilayer films at interfaces of each layer, *J. Vac. Sci. Technol.*, B21, 2, p. 785.

26. Miyake, S. (2005). Ttribology of carbonitride and boron nitride films and its application to nanoprocessing, *Thin Solid Films*, 493, 1–2, p. 160.
27. Miyake, S. (2004). Super-high density atomic force microscopy, *J. Sur. Finishing Soc. Jpn.*, 55, 12, p. 858 (in Japanese).
28. Miyake, S., and Asano, M. (2000). Nanometer scale mechanical processing of muscovite mica and application to high density memory, *IEEJ Trans. Sensors Micromach.*, 120-E, 1, p. 8 (in Japanese).
29. Miyake, S., and Matsuzaki, K. (2001). Realization of nanometer-scale processing by fluorocarbon plasma treatment of polycarbonate and its potential application to high density memory, *IEEJ Trans. Sensors Micromach.*, 121-E, 10, p. 564 (in Japanese).
30. Miyake, S., and Asano, M. (2000). Nanoprocessing by hard film coated tips and movement of processed nanoslider supported by van der Waals Forces, *J. Jpn. Soc. Precision Eng.*, 66, 8, p. 1275 (in Japanese).
31. Miyake, S., Otake, T., and Asano, M. (1999). Mechanical processing of standard rulers with one-nanometer depth of muscovite mica using an atomic force microscopy, *J. Jpn. Soc. Precision Eng.*, 65, 4, p. 570 (in Japanese).

Chapter 10

Nanomaterial Integration on CMOS Platform

Sumita Santra,[a] Prasanta Kumar Guha,[b] and Florin Udrea[a]
[a] *Electrical Engineering Department, University of Cambridge, CAPE Building, 9 J J Thomson Avenue, Cambridge CB3 0FA, UK*
[b] *Electronics and Electrical Communication Engineering Department, Indian Institute of Technology, Kharagpur 721302, India*
sumita.santra@gmail.com, pkguha@ece.iitkgp.ernet.in, fu10000@cam.ac.uk

10.1 Introduction

Nano (10^{-9} m length scale) materials have been around for more than two decades. And now microelectronics research has entered into the nano regime because of continuous demand for higher speed and higher packing density of transistors in integrated electronic devices. To meet this challenge, worldwide VLSI (Very Large Scale Integration) research both in academia and in the industry is concentrating on developing conventional (silicon MOSFETs available from many commercial foundries are already in nano-dimension) and unconventional (e.g., nano-tube, nano-pillars, nano-particles, nano-rods, etc.) nanostructures. Recently,

Vistas in Nanofabrication
Edited by Faiz Rahman

ISBN 978-981-4364-56-0 (Hardcover), 978-981-4364-57-7 (eBook)
www.panstanford.com

efforts are being made by some research groups to integrate these unconventional nanostructures with existing Complementary Metal Oxide Semiconductor (CMOS) Platform to combine the advantages of matured microelectronics technology with several superior properties offered by nanostructure materials [1]. This chapter will briefly discuss the present fabrication technologies for developing nanostructures and then mainly concentrate on the challenges in integrating these structures on CMOS Platform.

10.2 Nanostructure Fabrication

Several methods have been developed over the years to realize different nanostructures. Fabrication approaches are mainly based on either top-down (making smaller and smaller structure through etching of the substrate/bulk material as exemplified by the microelectronics industry) or bottom-up (starting from atoms or molecules to build nanostructures) methods. Popular techniques of growing nanomaterials include physical vapour deposition (PVD), chemical vapour deposition (CVD), sol-gel technique, hydrothermal methods, electron beam induced deposition techniques, and so on. Different recipes based on temperature, vapour pressure, nature of chemicals and their pH levels, duration of particular chemical treatments, thickness and nature of catalyst used, and many more dictate the nature and morphologies of the nanostructures formed. Often post-processing treatments (e.g., annealing at high temperature) are the key to make the material defect free, reliable, and useful for particular applications.

10.3 CMOS and Nanomaterials

Ingenious advancements for the last forty years in CMOS technology have been the main foundation for the sustained development of modern electronics and integrated circuits (IC). However, because of the demand for continuous miniaturization of device dimensions, it will soon be necessary to come up with some ground breaking technological inventions to continue the progress in IC technology. There are many experts who believe that nanomaterials (particularly carbon nanotubes) are the key to overcoming this challenge as these have the potential to replace existing CMOS technology. Although significant improvements have been made to synthesize nanomaterials efficiently and to locate them precisely on a substrate, there are several issues which still need to be solved before these unconventional nanomaterials can be combined with existing state-of-the-art CMOS technology. One of the main barriers is the lack of technology to realize nanostructures at wafer level and their unreliable performance as electronic devices compared to present-day silicon devices. So probably a more realistic approach will be to integrate nanomaterial on existing CMOS Platform in order to take advantage of both technologies.

10.3.1 Advantages of Integration

- CMOS is a well-established technology; it is possible to fabricate reliable, reproducible devices at wafer level.
- It is possible to develop intelligent electronic circuits (for various purposes) on a single chip with this technology.
- Because of batch fabrication, CMOS is a very low cost technology.

Hence, if one is able to integrate nanomaterials on a CMOS platform (rather than trying to just develop MOSFET devices out of these unconventional structure which might one day replace silicon) [2–7], both the technologies can complement each other. However, it is very challenging to integrate nanomaterials directly on CMOS Platform.

10.3.2 Integration Challenges

- A typical CMOS foundry will not allow any nanomaterial on their wafers before or during the process run because this can contaminate their system and also degrade the performance of ICs. Hence pre and intermediate deposition/growth of nanomaterials on CMOS wafer is very difficult. Thus, nanomaterial growth needs to be done as a post CMOS process.
- Very high temperatures (>500°C) are often required to

synthesize and anneal nanomaterials. This can be detrimental for other on-chip devices. Prolonged heat treatments above 400°C can cause reliability issues on the metal tracks because of aluminium electromigration.

- Use of harsh chemicals (required for nanomaterial synthesis) can damage semiconductor devices.

10.4 Nanomaterial Synthesis on CMOS Platform

Novel methods are being developed to synthesize nanostructures on CMOS wafers because of the above-mentioned constraints. These CMOS-friendly synthesis procedures should use much lower temperatures (to avoid any aluminium electromigration) and also utilize chemicals which are resistant to silicon oxide, silicon nitride, and aluminium.

10.4.1 Hydrothermal Method

This is a very popular method to form nanostructures without using very high temperatures and harsh environments. Although one needs to take care with the chemicals, which should not affect the CMOS substrate and the performance of other on-chip circuits. This approach is useful for developing almost any metal oxide nanostructure (e.g., zinc oxide nanowires (ZnO NWs), tin oxide nanorods, etc.). Zinc oxide nanowires are grown on a CMOS substrate as shown in Fig. 10.1 [8]. In this example, a thin ZnO seed layer was sputter-deposited (~5 nm) on selective areas of CMOS devices and spatially-defined using a metal shadow mask. These devices were then dipped in an equimolar (25 mM) aqueous solution of zinc nitrate hexahydrate ($Zn(NO_3)_2$ $6H_2O$, Sigma–Aldrich) and hexamethylenetetramine (HMTA, Sigma Aldrich) and were kept at 90°C for 2 hours. The devices were removed from the solution at the end of the growth, rinsed in de-ionised water, and dried under nitrogen flow. The advantage of this method is that ZnO NWs can be simultaneously grown on more than one device and, hence, can be extended to provide an inexpensive approach for wafer level fabrication.

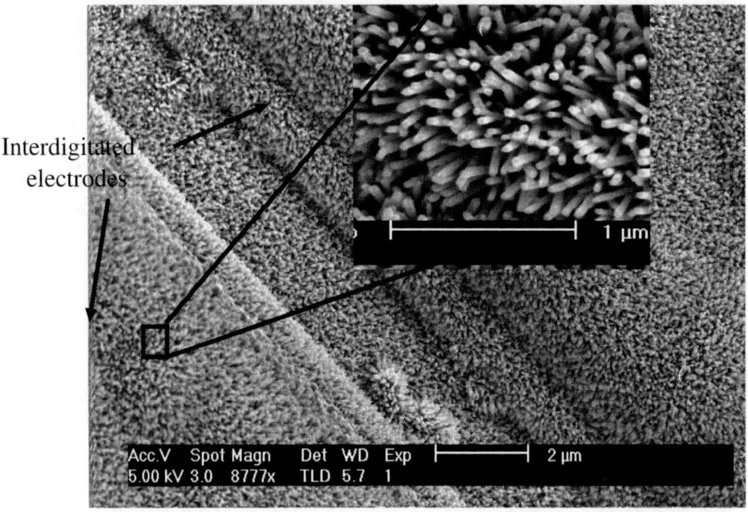

Figure 10.1 Scanning electron microscope (SEM) images of the ZnO NWs on interdigitated electrodes. Inset shows a higher magnification view.

10.4.2 Local Growth Technique

This is a relatively recent technique for growing nanostructures on specified areas of a substrate by a CVD approach. Since the catalyst is deposited on the entire chip and the vapour phase reactant is present everywhere within the chamber, the specific location of nanostructure growth has to be controlled by specific area heating (instead of heating the whole chip/substrate that will damage the on-chip electronics). Micro-heaters made of high-temperature metal/silicon/polysilicon can be used to locally raise the temperature of the catalyst which then forms small islands and then nanostructures while reacting with chemical vapours like in the conventional CVD approach. The local growth technique is a power efficient process (compared to heating the whole substrate), and it is also significantly faster because of the lower thermal mass. Englander *et al.* were one of the first to demonstrate this site-specific synthesis of nanostructures (silicon nanowire and carbon nanotubes) by using resistive heating in silicon/polysilicon suspended microbridges [9]. The silicon nanowires obtained (by localized joule heating for 15 min) are 30–80 nm in diameter and up to 5 mm in length, as shown in Fig. 10.2a,b,c. The pictures also clearly show location-dependent growth patterns as the result of a non-uniform temperature distribution on the microbridge. Later Jungen *et al.*, Sunden *et al.*, and some others reported similar local growth techniques for developing nanostructures [10,11]. However, the High Voltage Microelectronics and Sensors group [1] from the University of Cambridge was the first to demonstrate the growth of CNTs (both multiwall and single wall) on a fully processed CMOS wafer using this technique [12,13]. The micro-heater was made of tungsten (thus, avoiding any electromigration which is a cause for concern in the case of conventional CMOS metal, i.e., aluminium) and embedded in an oxide/nitride power-efficient ultrathin membrane. The structure (shown in Fig. 10.3) confines the heating to only the heater area, and, thus, there is no problem with integrating electronics on the same chip along with the fabricated nanostructure. The group went further and proposed a concept for extending their local growth technique to grow CNTs on specified areas at wafer level [14]. The basic scheme is shown in Fig. 10.4. The concept is that all the micro-hotplates, spread across the wafer, are connected to each other via tracks in the scribe lanes between chips. This allows the heating of each micro-hotplate on the wafer at the same time, and, hence, local growth of CNTs on each micro-hotplate takes place simultaneously.

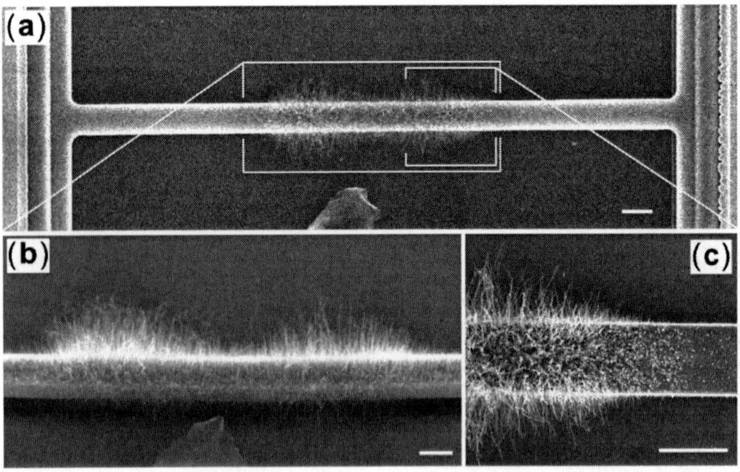

Figure 10.2 (a) SEM images of silicon nanowires that were grown by the local growth method (b) and (c) Close-up view of the silicon nanowires. Reprinted with permission from American Institute of Physics [9].

After CNT growth, the dicing of the wafer removes the connection between the chips, leaving each chip with its own micro-hotplate and CNTs. As a proof of concept, they showed simultaneous growth of identical CNTs at adjacent chips (shown in Fig. 10.5). The on-chip growth of multiwall and single wall CNTs are shown in Figs. 10.6 and 10.7, respectively.

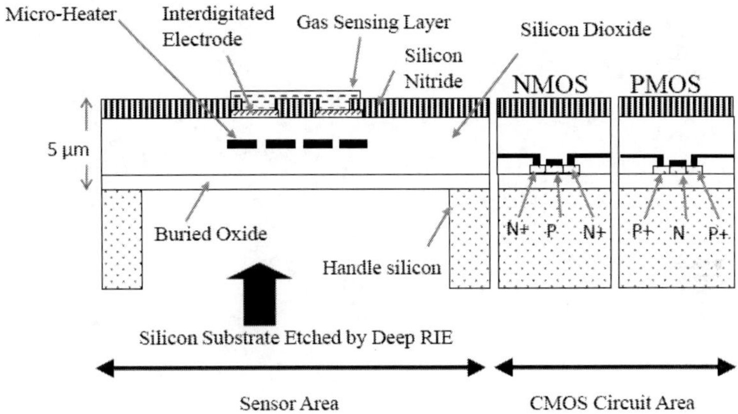

Figure 10.3 Cross-sectional view of CMOS micro-hotplate.

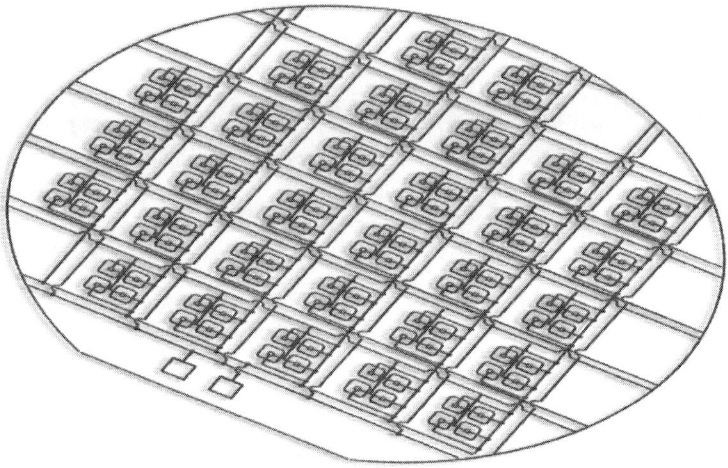

Figure 10.4 A schematic diagram of wafer-level carbon nanotubes growth. Reprinted with permission from Institute of Physics [14].

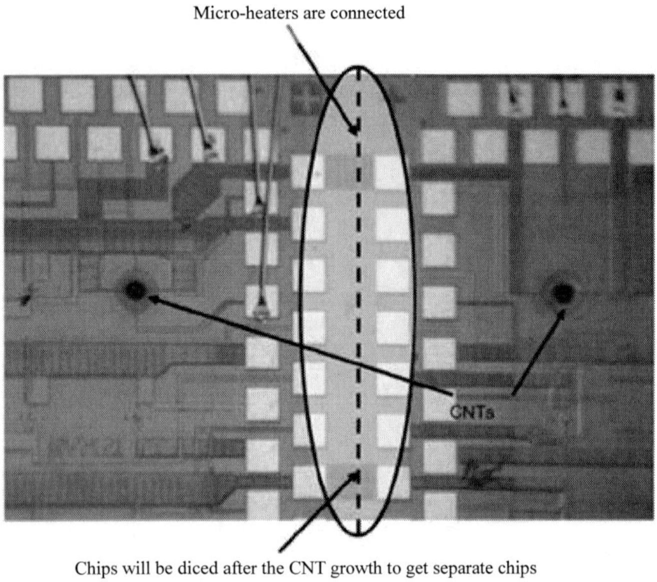

Figure 10.5 Optical microscope picture of CNTs which were locally grown simultaneously on two micro-hotplates at adjacent chips. Reprinted with permission from Institute of Physics [14].

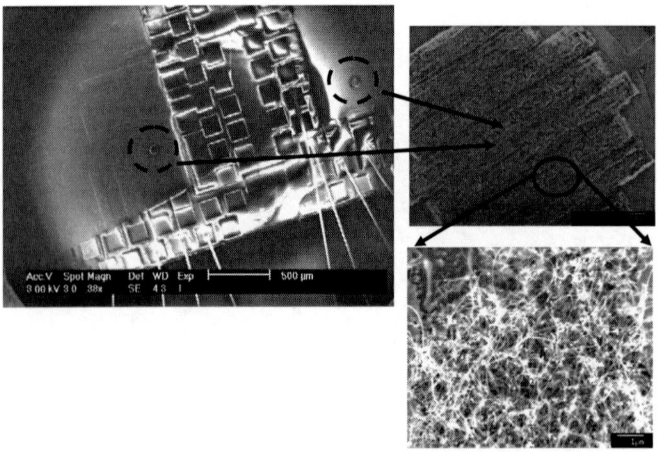

Figure 10.6 SEM picture of the MWCNTs which were locally grown simultaneously on two micro-heaters at adjacent chips. MWCNTs were grown on the interdigitated electrodes. Close-up view of the spaghetti-like MWCNTs on the interdigitated electrode. Reprinted with permission from Institute of Physics [14].

Figure 10.7 SEM image of the SWCNTs on micro-hotplate grown by the local growth technique.

10.4.3 Flame Spray Pyrolysis

Liquid feed-Flame pyrolysis is a one-step process by which a metal precursor(s), dissolved usually in an organic solvent, is sprayed with an oxidizing gas into a flame zone where the droplets are combusted and the precursor(s) converted into nano-sized metal or metal-oxides particles, depending on the metal and the operating conditions. Flame spray pyrolysis (FSP) is a very promising technique for nano-sensor material fabrication since it enables primary particle and crystal size control, which is important to tailor sensitivity, as well as the controlled *in situ* deposition of noble metal clusters. Another important feature is that FSP can be utilized for *in situ* direct deposition of porous sensitive layers onto diverse substrates. A typical FSP set-up described in [15] is shown in Fig. 10.8. The substrate wafer is clamped between the shadow mask and the chilled support plate and mounted upside down above the deposition flame. During deposition, the wafer temperature was held at 120°C. Kühne *et al.* have already reported wafer level post-CMOS FSP deposition of SnO_2/Pt gas sensitive layers on microsensors, as shown in Fig. 10.9 [15].

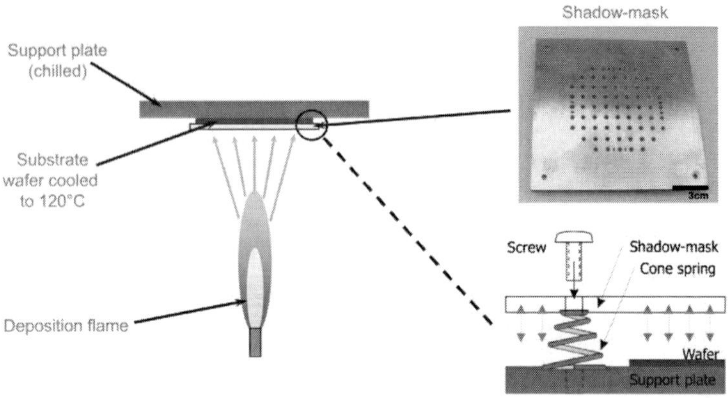

Figure 10.8 Illustration of the FSP deposition set-up including a picture of the shadow mask and a sketch of the substrate clamping mechanism. Reprinted with permission from Institute of Physics [15].

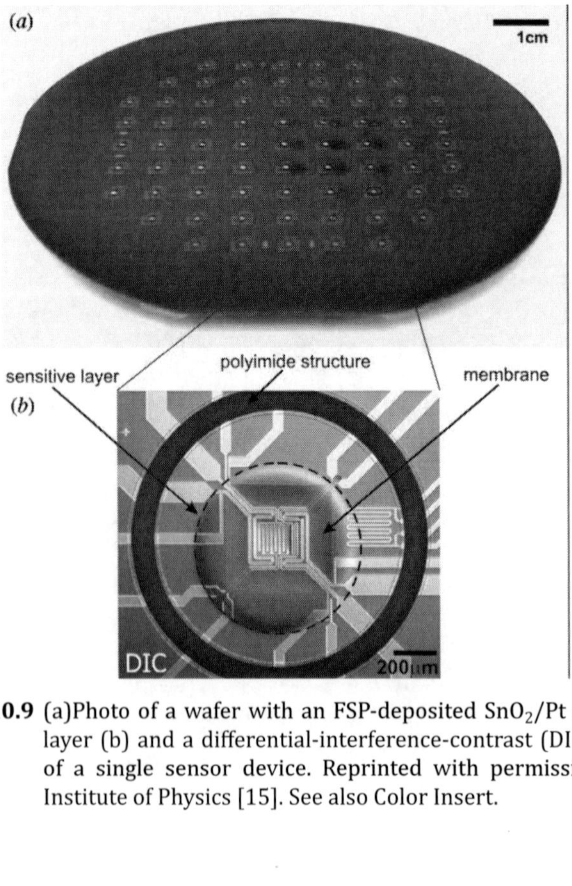

Figure 10.9 (a)Photo of a wafer with an FSP-deposited SnO_2/Pt sensitive layer (b) and a differential-interference-contrast (DIC) image of a single sensor device. Reprinted with permission from Institute of Physics [15]. See also Color Insert.

10.4.4 Aerosol-Assisted Chemical Vapour Deposition

Chemical vapour deposition techniques such as aerosol-assisted chemical vapour deposition (AACVD) can be used to grow nanomaterials on CMOS platform if the process temperature can be reduced. This is a CVD process in which the precursors are transported to the substrate by means of a liquid/gas aerosol, which can be generated ultrasonically. This technique is suitable for use with non-volatile precursors. In Fig. 10.10, scanning electron microscope picture of tungsten oxide (WO_3) film deposited by AACVD is shown. The average particle size of around 500 nm of WO_3 was deposited from AACVD reaction of $[^nBu_4N]_2[W_6O_{19}]$ at 550°C [16].

Figure 10.10 SEM micrographs of tungsten oxide films deposited from AACVD reactions of $[^nBu_4N]_2[W_6O_{19}]$ at 550°C. Reprinted with permission from The Royal Society of Chemistry [16].

10.4.5 Inkjet Printing/Spray Coating

Sometimes researchers use techniques like inkjet printing/ spray coating or dipping in the solution to deposit nanomaterials on a surface. This is usually done with commercially available nanomaterials, so one can avoid harsh environments which are often necessary to grow nanomaterials. Printers with micrometer precision and automated programmable deposition facilities are already available in the market [17,18]. Even though these methods are low cost and CMOS friendly, dispersing nanomaterials in a carrier is often difficult due to strong van der Waals force. One might, therefore, have to use a high-speed sonicator for hours to avoid nozzle clogging in the printer or spray coater. For dip coating, one needs to cover bond pads to avoid any chemical contaminations. It is also necessary to mention that often these methods, particularly deposition of nanomaterial with inkjet printing, do not require conventional lithography approaches (i.e., steps like deposition of photoresist, exposure of ultra violet light, lift off, etc.) or use of some masking process (e.g., shadow mask) to grow or deposit nanomaterials on predefined areas of the chip. Furthermore, the use of these techniques is restricted to low volumes (unless it is automated) and, hence, negates the use of CMOS technology [19].

10.5 Conclusion

In this chapter, we have discussed the integration of nanomaterials on CMOS platform. This integration is important to combine the benefits from both technologies. Researchers have been developing different methods to synthesis nanomaterials on precisely-predefined areas of a substrate. However, to grow them on a fully-processed CMOS wafer needs methods which require low process temperature (<400°C) and CMOS-friendly chemicals. Here, we have described some such methods (like hydrothermal, local growth, FSP, and inkjet printing) which can be very useful to combine the two technologies together without degrading their separate features.

References

1. http://www-g.eng.cam.ac.uk/hvm/index.html.
2. Kong, J., Franklin, N.R., Zhou, C.W., Chapline, M.G., Peng, S., Cho, K.J., and Dai, H.J. (2000) Nanotube molecular wires as chemical sensors, *Science*, **287(5453)**, pp. 622–625.
3. Stampfer, C., Helbling, T., Obergfell, D., Schoberle, B., Tripp, M.K., Jungen, A., Roth, S., Bright, V.M., and Hierold, C. (2006) Fabrication of single-walled carbon-nanotube-based pressure sensors, *Nano Lett*, **6(2)**, pp. 233–237.

4. Dai, H.J., Hafner, J.H., Rinzler, A.G., Colbert, D.T., and Smalley, R.E. (1996) Nanotubes as nanoprobes in scanning probe microscopy, *Nature*, **384(6605)**, pp. 147–150.

5. Kim, P. and Lieber, C.M. (1999) Nanotube nanotweezers, *Science*, **286(5447)**, pp. 2148–2150.

6. Jensen, K., Weldon, J., Garcia, H., and Zettl, A., (2007) Nanotube radio, *Nano Lett*, **7(11)**, pp. 3508–3511.

7. Snow, E.S., Perkins, F.K., Houser, E.J., Badescu, S.C., and Reinecke, T.L. (2005) Chemical detection with a single-walled carbon nanotube capacitor, *Science*, **307(5717)**, pp. 1942–1945.

8. Santra, S., Guha, P.K., Ali, S.Z., Hiralal, P., Unalan, H.E., Covington, J.A., Amaratunga, G.A.J., Milne, W.I., Gardner, J.W., and Udrea, F. (2010) ZnO nanowires grown on SOI CMOS substrate for ethanol sensing, *Sens Actuators B-Chem*, **146(2)**, pp. 559–565.

9. Englander, O., Christensen, D., and Lin, L.W. (2003) Local synthesis of silicon nanowires and carbon nanotubes on microbridges, *Appl Phys Lett*, **82(26)**, pp. 4797–4799.

10. Jungen, A., Stampfer, C., Tonteling, M., Schiesser, S., Sarangi, D. and Hierold, C. (2005) Localized and CMOS compatible growth of carbon nanotubes on a 3 × 3 μm^2 microheater spot, *Dig Tech Papers Trans*, **1,** pp. 93 –96.

11. Sunden, E.O., Wright, T.L., Lee, J., King, W.P., and Graham, S. (2006) Room-temperature chemical vapor deposition and mass detection on a heated atomic force microscope cantilever, *Appl Phys Lett*, **88(3)**, pp. 3.

12. Haque, M.S., Ali, S.Z., Guha, P.K., Oei, S.P., Park, J., Maeng, S., Teo, K.B.K., Udrea, F., and Milne, W.I. (2008) Growth of carbon nanotubes on fully processed silicon-on-insulator CMOS substrates, *J Nanosci Nanotechnol*, **8(11)**, pp. 5667–5672.

13. Haque, M.S., Teo, K.B.K., Rupensinghe, N.L., Ali, S.Z., Haneef, I., Maeng, S., Park, J., Udrea, F., and Milne, A.I. (2008) On-chip deposition of carbon nanotubes using CMOS microhotplates, *Nanotechnology*, **19(2)**, pp. 5.

14. Santra, S., Ali, S.Z., Guha, P.K., Zhong, G.F., Robertson, J., Covington, J.A., Milne, W.I., Gardner, J.W., and Udrea, F. (2010) Post-CMOS wafer level growth of carbon nanotubes for low-cost microsensors-a proof of concept, *Nanotechnology*, **21(48)**, pp. 7.

15. Kuhne, S., Graf, M., Tricoli, A., Mayer, F., Pratsinis, S.E., and Hierlemann, A. (2008) Wafer-level flame-spray-pyrolysis deposition of gas-sensitive layers on microsensors, *J Micromechan Microengin*, **18(3)**, pp. 10.

16. Ashraf, S., Blackman, C.S., Palgrave, R.G., and Parkin, I.P. (2007) Aerosol-assisted chemical vapour deposition of WO_3 thin films using polyoxometallate precursors and their gas sensing properties, *J Mater Chem*, **17(11)**, pp. 1063–1070.

17. http://www.altatech-sc.com/en/index.php.

18. http://www.tesscorn.com/nanotechnology.htm.

19. Udrea, F., Santra, S., Guha, P. K., Ali, S. Z., Covington, J. A., Milne, W. I., Gardner, J. W., and Maengc, S. (2009) Nanotubes and nanorods on CMOS substrates for gas sensing, in *Proc 13th Intl Symposium Olfaction Electron Nose*, **1137**, pp. 19–26.

Chapter 11

Focused Ion Beam Fabrication of Metallic Nanostructures

Michal Urbánek* **and Tomáš Šikola**
Central European Institute of Technology, CEITEC BUT,
Brno University of Technology, Technická 10, Brno, CZ-616 00, Czech Republic
Institute of Physical Engineering, Faculty of Mechanical Engineering,
Brno University of Technology, Technická 2, Brno, CZ-616 69, Czech Republic
*urbanek@fme.vutbr.cz

This chapter describes the use of focused ion beam (FIB) system for the nanofabrication of nanostructures from sputter-deposited polycrystalline metallic thin films. FIB milling of polycrystalline metallic thin films significantly differs from the milling of single crystalline and amorphous materials, and great care must be taken to overcome inherent structural limitations in the milled thin film when fabricating fine nanostructures with sub 100 nm resolution.

11.1 Introduction

Nanopatterning of metallic thin films is becoming increasingly important due to recent developments in nanoelectronics and nanophotonics. There are important applications of such techniques

Vistas in Nanofabrication
Edited by Faiz Rahman

ISBN 978-981-4364-56-0 (Hardcover), 978-981-4364-57-7 (eBook)
www.panstanford.com

in spintronics [1] and plasmonics [2], where information is stored and/or transferred by using nanoscale metallic elements. Such structures are usually fabricated by electron beam lithography while patterning by FIB is a much less frequent technique even though it possesses a lot of advantages, especially in research and rapid prototyping applications [3,4].

The main advantage of this method compared to usual lithography techniques is speed and simplicity of the maskless one-step process based on direct ion-beam writing, i.e., transferring a pattern by direct impingement of the FIB onto the sample. FIB direct writing can combine a series of techniques such as milling, implantation, ion-induced deposition (see Fig. 11.5), and ion-assisted etching.

The main drawback in the patterning of nanostructures from metallic thin films by FIB milling is residual roughness, which is also transferred onto the edges of milled patterns and limits the quality and resolution of fabricated nanostructures. The key issue in nanofabrication by FIB is a proper choice of ion beam parameters such as ion beam current and its profile (spot size), scan routine, and speed over the sample, among others.

11.2 Focused Ion Beam Systems

A two-beam FIB system combines imaging capabilities of the SEM column with the direct writing and imaging capabilities of a FIB column. Today's FIB systems utilize a liquid-metal ion source to produce ions, typically Ga^+. The ions are extracted and focused into a beam by an electrostatic field. The beam then passes through an aperture and is finally scanned over the sample surface. The ions can then be used for etching or milling the surface, as well as for its imaging. Due to the ions' higher mass and larger size, the interaction of ions with the sample significantly differs from interaction with much lighter electrons. Ions cannot penetrate deep into the material, the typical penetration depth in metals is approximately 20 nm for 30 keV Ga^+ ions, whereas the penetration depth for 30 keV electrons is approximately 2 μm. Also, one ion typically produces one secondary electron and five sputtered atoms (Ga^+, 30 keV into Cu), whereas one electron roughly produces 0.5 secondary electron and no sputtered atom. However, due to the destructive nature of the ion beam, electrons generated in a SEM column and not ions are preferentially used for simultaneous non-destructive high-resolution sample observation. Other typical components of an FIB system are a gas injection system (GIS), which is used for FIB-induced deposition of materials from metalorganic gases, and nanomanipulators for nanostructure probing and manipulation (Fig. 11.1).

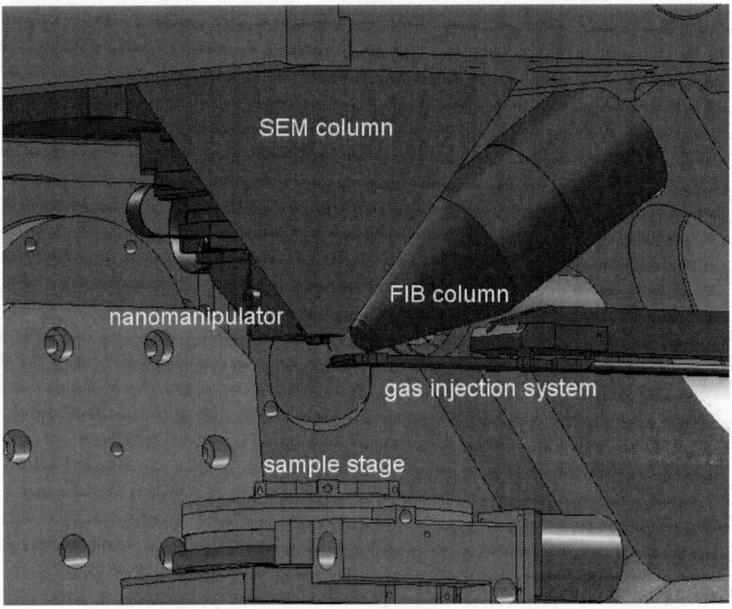

Figure 11.1 View inside a FIB system chamber (TESCAN Lyra). Electron and ion beams meet at a coincidence point which is located 9 mm under the SEM column. Other components such as GIS and nanomanipulators are also located around this point. A computer-controlled stage allows eucentric tilting of the sample against the FIB or SEM columns for undistorted observation.

11.3 Milling of Metallic Thin Films

When 30 keV Ga^+ ions hit the surface of a metallic film, they penetrate to a depth of approximately 5–40 nm. On their way through the film, the ions lose their energy mostly due to interactions with target atoms and form collision cascades [5]. If an atom is provided with sufficient energy through such collisions, it can overcome the surface binding energy and escape from the sample, i.e., it can be sputtered away. The ratio of atoms sputtered per one incident ion at given ion beam energy is a statistical property of the material and is called the sputter yield (atoms/ion). Ideally, there should be a linear relationship between the ion dose and the sputtered volume. However, in the sputtering process several additional parameters and their associated effects are involved and must be considered:

(i) **Variance in the local angle of incidence**. The sputter yield increases with the incidence angle (measured from the surface normal) up to 80° and then rapidly drops. When milling inhomogeneous and rough surfaces, the varying local incidence angle causes uneven milling rate over the sample surface. This results in further evolution of surface morphology and leads to surface rippling [6,7].

(ii) **Channelling**. When milling polycrystalline thin films, the sputter rate depends on their crystallographic orientation with respect to the incident ion beam. At certain orientations, ions easily channel deep into the material, thus making it less probable that the collision cascade bounces out atoms. This causes a relative decrease of the sputter yield in grains where the channelling can occur in comparison with grains where channelling is less dominant [8].

(iii) **Ion-induced grain growth (recrystallization)**. Ion irradiation leads to grain growth in polycrystalline thin films [9]. The grains oriented favourably for channelling can increase in size under 30 keV FIB bombardment [10].

(iv) **Redeposition**. As the sputtered atoms are not in their thermodynamic equilibrium state, they tend to condense back into the solid phase after their collision with any nearby surface. The redeposition can be affected by the scanning strategy (one pass vs. multiple passes), scanning pattern (e.g., raster vs. spiral), shape (aspect ratio) of the milled structures [11], dynamics of the ejected atoms, and the sticking coefficient of the target material.

Metallic thin films prepared by sputter deposition exhibit polycrystalline structure which can be imaged by FIB due to the channelling effect (Fig. 11.2a). The grains where the crystallographic orientation allows channelling are black, and the grains where no channelling occurs are imaged as white. The channelling causes uneven sputtering over the sample surface, which results in the development of a rough surface. After the application of sufficiently high ion dose, the grains become visible also in SEM images (Fig. 11.2b) due to topographic contrast.

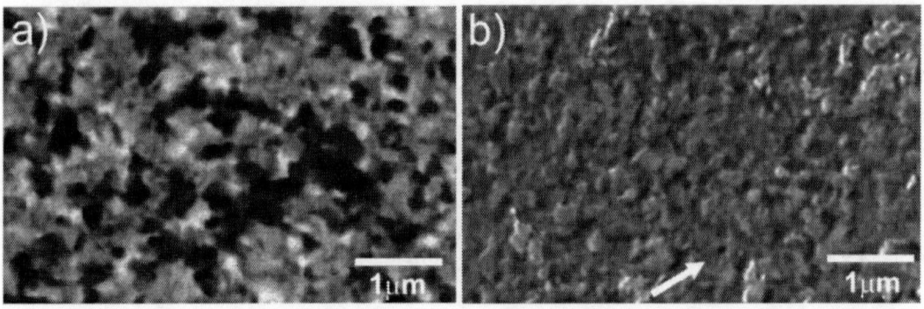

Figure 11.2 (a) FIB image of a virgin metallic thin film (cobalt, imaged using SE detector) showing a channelling contrast caused by different crystallographic orientations of grains. (b) SEM image of the same film after irradiated by an ion dose of 0.024 nC/μm^2. The arrow points to the dark hole with round edges indicating reaching the Co/Si interface [12].

11.3.1 Evolution of the Milling Process for Metallic Thin Film on Silicon Substrate

The evolution of the milling process is described with the example of a 30 nm Co thin film on a silicon substrate [12], however, similar behaviour was also observed in the milling of other metallic thin films (Co, Cu, Au, Al, Pt, Permalloy) and multilayer structures (Permalloy/Cu/Co spin-valves) prepared by sputter deposition and having thicknesses ranging from 5 to 100 nm.

A 30 nm Co thin film on a Si substrate was irradiated step-by-step with increasing ion dose, and the bottom of the milled surface was subsequently measured by the AFM. In the first stages of milling, where only the cobalt thin film was affected, the sputter rate decreased at the grains where ions channelled deeper under the surface and the resulting milled depth was lower. The sputter yield at grains where no channelling occurred was approximately two times higher (Fig. 11.3a,b). The difference in the sputter yields for various grain orientations caused the topography of the milled surface to partially replicate the grain structure of the thin film.

As the bottom of the milled structures approached the Co/Si interface, the surface topography, which originally replicated the cobalt grains, changed. The grain edges rounded and a random ripple pattern evolved (Fig. 11.3c,d).

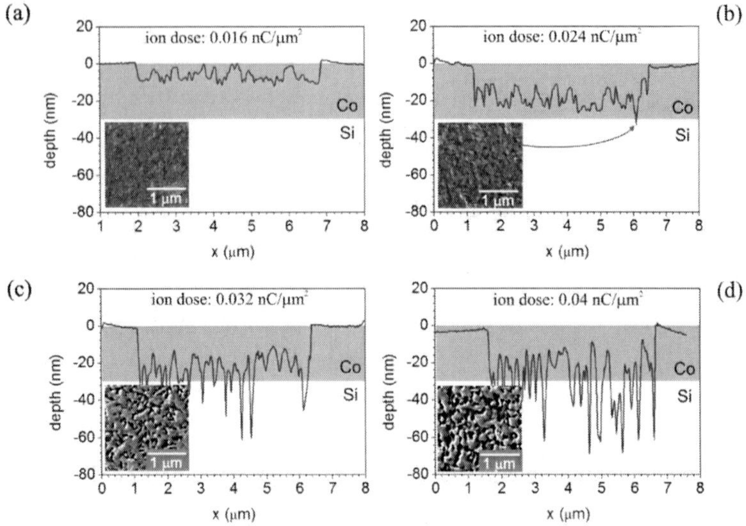

Figure 11.3 Milling a 30 nm Co thin film by 30 keV Ga^+ ions, at normal incidence angle. AFM profiles together with corresponding SEM images of the milled surface after ion irradiation with increasing ion dose are shown. (a) Only the thin film is affected — the roughness profile follows the grain distribution in the thin film. (b) Co/Si interface is reached, the first evidence of the ripples is shown by the arrow. (c, d) Further evolution of the ripple structure [12]. See also Color Insert.

The origin of ripple formation in silicon is a topographic variation of the remains of the top cobalt layer. A monocrystalline silicon substrate does not exhibit variable sputter yield caused by channelling through the grains. However, the topography of the sputtered cobalt layer significantly influenced the milling of the silicon substrate. The milling was driven by the following effects: valleys and pits were milled faster due to the focusing effects of their sidewalls — most of the Ga^+ ions were reflected forward when they hit almost vertical sidewalls, i.e., they were aimed towards the bottom of the milled structures. This created deep narrow holes, as can be seen in Fig. 11.3c,d. Another effect leading to a random ripple pattern is the angular dependence of the sputter yield. The sidewalls of the milled structures, which were not steep enough to reflect Ga^+ ions (for sidewall angles smaller than 80°), were milled faster than the elevated sites where ions hit the surface at right angle. The redeposition also played a role here as the atoms sputtered from the trenches and pits redeposited at the edges and further slowed down the milling of the elevated areas. All these effects were responsible for an increase in roughness at the polycrystalline metal/monocrystalline silicon interface.

11.3.2 Estimation of the Ion Dose Needed for Thin Film Removal

When milling a thin, film it is often difficult to find the point when the substrate is reached. This point can be easily deduced from SEM images by observation of the change of the milling mode. As dark holes with round edges in-between the sharp grains start to appear, they provide the first suggestion of a change of the milling mode at the polycrystalline metal/monocrystalline silicon interface. This can be used for ion dose and milling time estimation. This effect is shown in Figs. 11.2b and 11.3b.

11.4 Influence of Ion Beam Parameters

Ion beam parameters are critical in achieving optimum results in FIB machining. In the case of milling of polycrystalline thin films, the main task is to find the optimum values for suppressing the effects of surface roughening caused by ion channelling. The parameters which can be controlled for this are ion beam current, the time the beam rests in one position (dwell time), relative distance between beam positions (overlap), and ion beam incidence angle (Fig. 11.4).

Ion-beam current directly influences the spot size and the milling speed. Both characteristics increase with increasing the current. For the fabrication of nanostructures with ultimate resolution, the use of high currents (>100 pA) is not feasible as it results in large spot sizes. However, there is no need to use lowest currents either (e.g., imaging mode, beam current ~1 pA, spot size ~7 nm), because slightly higher currents with somewhat larger spot sizes (e.g., 36 pA, spot size ~15 nm) still do not result in an increase of roughness for depths smaller than the thickness of the thin film (Fig. 11.5). Sputter yields at distinct grains do not depend on the ion-beam current, and the increased spot sizes at currents up to 50 pA are still well below the size of the thin film grains. When the bottom of the milled structures reaches the thin film/substrate interface, the role of the grains in the milling process is suppressed and the higher currents can produce rougher surfaces, however, at this point the milling should be stopped as the thin film is removed and the structure is fabricated.

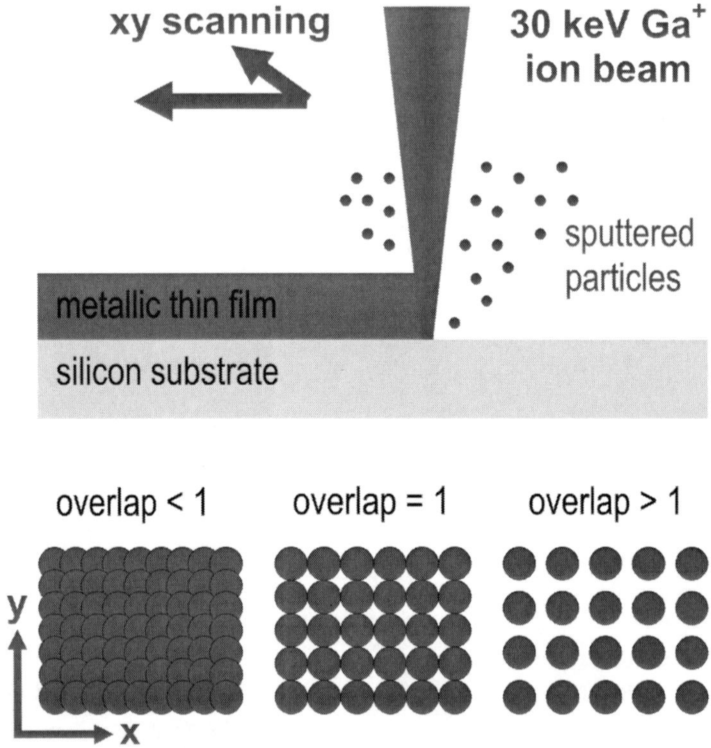

Figure 11.4 Schematic of the FIB milling process. A 30 keV Ga^+ ion beam is focused on the spot (size ≥ 7 nm) and is scanned across the selected areas of the surface. The beam current, duration at each impact point (dwell time), distance between the impact spots (overlap), and the angle of incidence of the ion beam can all be independently controlled.

Another parameter influencing the milling process is the dwell time (the time defining how long the ion beam stops at each pixel). This parameter does not directly influence the roughness; however, it significantly influences the milling process in two different ways: the first effect is an increase in the sputter yield for longer dwell times. After the initial reduction of the sputter yield due to the effect of swelling [13], the sputter yield for the longer dwell times can increase by 30%. This value is valid for perpendicular ion beam incidence, whereas for a tilted ion beam angle the difference diminishes, and for an ion beam incidence angle of 55°, the difference is only about 10%. This variation in the sputter yield is due to the formation of non-vertical sidewalls caused by slower scanning by the Gaussian-shaped ion beam. This leads to a dwell-time dependence on the local angle of incidence [14]. It is often reported [15] that using slow single scans (long dwell times) leads to the formation of structures with inclined bottoms. However, this was observed for high-aspect-ratio (1:10) micron-sized structures and has not been observed in low-aspect-ratio (1:1) nanostructures typically fabricated from metallic thin films.

Another effect which needs to be considered in selecting the dwell time is ion beam-induced recrystallization. For longer dwell times, the grains can change their size significantly — a grain oriented for channelling can grow by a factor of 10, and this effect needs to be taken into account when observing thin film behaviour under milling and estimating optimal beam parameters.

The roughness of the milled structure increases steeply for overlap ≥ 1. The biggest improvement in surface smoothness occurs when the overlap is approximately 0.7; further reduction of the overlap has little or no effect. At this overlap, the homogeneity of the ion dose areal distribution is 99% (assuming a Gaussian beam shape). These results correspond to studies made on silicon [11]. A decrease in the surface roughness is again more evident for larger depths where the substrate is involved, and the role of grains is suppressed.

The most influential parameter is the angle of incidence of the ion beam. The sputter yield increases with increasing the angle. At the angle of 55°, the sputter yield is 3.5× higher than at 0°. Such an increase of the sputter yield combined with an increase caused by longer dwell times can be used to minimize unwanted damage caused by collision cascades and Ga implantation. The reduction of the ion-dose required for the complete removal of the thin film is important especially in the fabrication of layered magnetic structures. The tail of the Gaussian beam extends to 100 nm in diameter even for the smallest spot sizes (<10 nm) [16] and can cause material intermixing at the borders of the milled structures. Typically, the ion doses needed to destroy the magnetic properties of, for instance, Co thin films are quite high, in the order of 10–12 C/μm^2 [16], which is above the dose introduced by the tail of the Gaussian beam during the removal of a 30 nm thin film. However, magnetic multilayers (i.e., spin-valve sensors) can be generally more sensitive due to the interlayer mixing caused by collision cascades [17].

Besides increasing the sputter yield, the angle of incidence of the ion beam has a strong influence on surface roughness. The roughness of the milled structures decreases significantly with decreasing ion-beam incidence angle. The effect of different sputter yields at different grains of the thin film, resulting in enhanced surface roughness, is suppressed due to shielding of the milled valleys and pits by surrounding protrusions. The protrusions are preferentially milled at their peaks due to the oblique ion-beam incidence. Although the lowest residual roughness can be achieved for angles up to 80°, due to practical reasons, the angle of choice is typically the angle between SEM and FIB columns (typically 50–55°, depending on the FIB system). At this angle, the tapering and undercutting of the structures is less pronounced and it is possible to simultaneously observe fabricated structures by the SEM at the normal angle. By milling the structures at a tilted angle, it is possible to substantially improve their quality, however, the tapering and undercutting of the structures milled by the inclined ion beam have to be taken into account especially in structures with relatively high aspect ratios.

11.5 Contacting the Structures

One of the biggest advantages of FIB systems is their ability to fabricate electrical contacts directly through FIB-induced deposition [18,19] and/or to provide *in situ* electrical measurements by nanomanipulators [20]. This makes the FIB technology suitable for fast, one-step-process fabrication and analysis of nanostructures for nanoelectronics. This is particularly advantageous in research and rapid prototyping applications. In Fig. 11.5, we show an example of a magnetic nanowire with platinum contacts used for magnetoresistance measurements. The fabrication of such a sample by FIB is much faster and easier than by using multistep e-beam lithography.

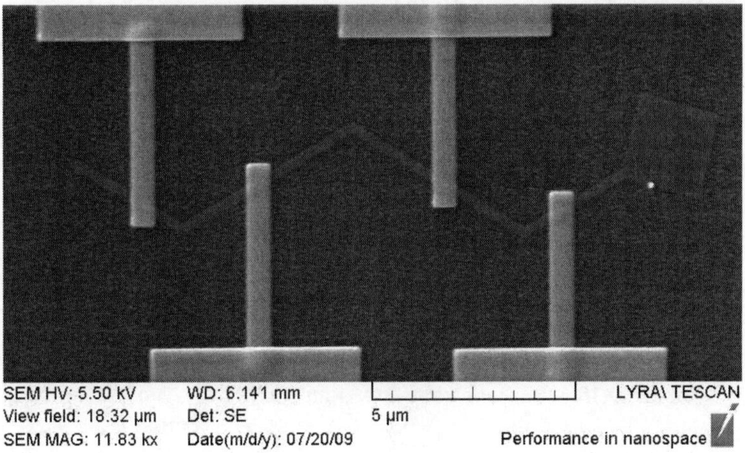

Figure 11.5 Magnetic nanowire of 200 nm width fabricated from a magnetic multilayer by FIB milling with platinum contacts deposited by FIB-induced deposition.

11.6 Summary

FIB milling of the sputtered metallic thin films significantly differs from milling of the monocrystalline materials. Surface roughening and transfer of the roughness to the edges of patterned structures limit the quality and resolution of fabricated nanostructures. The main reasons for the increase of roughness are different sputter yields for various crystallographic orientations of thin film grains resulting from ion channelling. The quality of the milled structures can be significantly improved by a proper choice of ion beam parameters. Long dwell time and oblique angle of incidence are the most important parameters for obtaining smooth surfaces. A further improvement of the structural quality can be obtained by considering the fact that the surface roughening is caused by channelling through the grains. This can be surpassed either indirectly by using more uniform thin films (epitaxial layers or layers deposited with an assisted ion beam) or directly by selective milling of the grains with a variable dwell time, i.e., by using the secondary electron signal as a feedback during the milling process.

References

1. Chappert, C., Fert, A. and van Dau, F. N., (2007) The emergence of spin electronics in data storage, *Nat. Mater.,* **6**, pp. 813–822.
2. Maier, A. S. and Atwater H. A., (2005) Plasmonics: localization and guiding of electromagnetic energy in metal/dielectric structures, *J. Appl. Phys.*, **98**, 011101.
3. Tseng, A. A., (2005) Recent developments in nanofabrication using focused ion beams, *Small,* **1,** pp. 924–939.
4. Gierak, J., (2009) Focused ion beam technology and ultimate applications, *Semicond. Sci. Technol.*, **24,** 043001.
5. Imanishi, N., (2007) Interactions of ions with matter, *Focused Ion Beam Systems: Basics and Applications (*ed. N Yayo) (Cambridge University Press, Cambridge), pp. 31–66.
6. MoberlyChan, W. J., Felter, T. E. and Wall, M. A., (2006), Surface rippling & ion etch yields of diamond using a focused ion beam: with or without enhanced chemistry, aspect ratio regulates ion etching, *Microsc. Today*, **14**, pp. 28–34.
7. Muñoz-García, J., Vázquez, L., Cuerno, R., Sánchez-García, J. A., Castro, M. and Gago, R., (2009), Self-organized surface nanopatterning by ion beam sputtering, *Towards Functional Nanomaterials* (ed. Wang, Z. M.) (Springer, New York).
8. Levi-Setti R., Fox, T., R. and Lam, K., (1983), Ion channeling effects in scanning ion microscopy with a 60 keV Ga^+ probe, *Nucl. Instrum. Methods,* **205,** pp. 299–309.
9. Attwater, H. A., Thompson, C. V. and Smith, H. I., (1988), Ion-bombardment-enhanced grain growth in germanium, silicon, and gold thin films, *J. Appl. Phys.,* **64** (5), pp. 2337–2353.
10. Spolenak, R., Sauter, L., Eberl, C., (2005), Reversible orientation-biased grain growth in thin metal films induced by a focused ion beam, *Scripta Mater.,* **53,** pp. 1291–1296.
11. Lipp, S., Frey, L., Lehrer, C., Frank, B., Demm, E., Ryssel, H., (1996), Investigations on the topology of structures milled and etched by focused ion beams, *J. Vac. Sci. Technol. B,* **14,** pp. 3996–3999.
12. Urbánek, M., Uhlíř, V., Bábor, P., Kolíbalová, E., Hrnčíř, T., Spousta, J., Šikola, T., (2010), Focused ion beam fabrication of spintronic nanostructures: an optimization of the milling process, *Nanotechnology,* **21**, 145304.
13. Lehrer, C., Frey, L., Petersen, S., Ryssel, H., (2001), Limitation of focused ion beam nanomachining, *J. Vac. Sci. Technol. B,* **19**, pp. 2533–2538.

14. Santamore, D., Edinger, K., Orloff, J., and Mengailis, J., (1997), Focused ion beam sputter yield change as a function of scan speed, *J. Vac. Sci. Technol. B,* **15,** pp. 2346–2349.

15. Adams, D. P. and Vasile, M. J., (2006), Accurate focused ion beam sculpting of silicon using a variable pixel dwell time approach, *J. Vac. Sci. Technol. B,* **24,** pp. 836–844.

16. Gierak, J., *et al.* (2005), Exploration of the ultimate patterning potential achievable with focused ion beams, *Microelectron. Eng.,* **78–79,** pp. 266–278.

17. Kerr, E., van Dijken, S., Langford, R. M., Coey, J. M. D., (2005), Effect of Ga^+ ion implantation on the magnetoresistive properties of spin valves, *J. Magn. Magn. Mater.,* **290–291,** pp. 124–126.

18. Vila, A., Hernandez-Ramirez, F., Rodriguez, J., Casals, O., Romano-Rodriguez, A., Morante, J. R., Abid, M. (2006), Fabrication of metallic contacts to nanometre-sized materials using a focused ion beam (FIB), *Mater. Sci. Eng. C,* **26**, pp. 1063–1066.

19. MoberlyChan, W. J., Adams, D. P., Aziz, M. J., Hobler, G. and Schenkel, T., (2007), Fundamentals of focused ion beam nanostructural processing: below, at, and above the surface, *MRS Bull.,* **32**, pp. 424–432.

20. Gupta, R. and Stallcup II, R. E., (2007), Introduction to *in situ* nanomanipulation for nanomaterials engineering, *Scanning Microscopy for Nanotechnology: Techniques and Applications* (ed. Zhou, W. and Wang, Z. L.) (Springer, New York), pp. 192–224.

Nanotechnology Showcase

The entries in this part of the book show the diverse structures and length scales that are encountered in micro- and nanofabrication. Here, nine research groups demonstrate interesting structures from the fields of materials science and device engineering. Each entry consists of one or more micrographs together with accompanying explanatory text and a list of references for further information.

Size-Tunable Silicon Nanopore Arrays

X. Chen and K. Jiang
Bio-medical and Micro Engineering Research Centre,
School of Mechanical Engineering, University of Birmingham, B15 2TT, UK
x.chen.2@bham.ac.uk, samchen@hotmail.co.uk

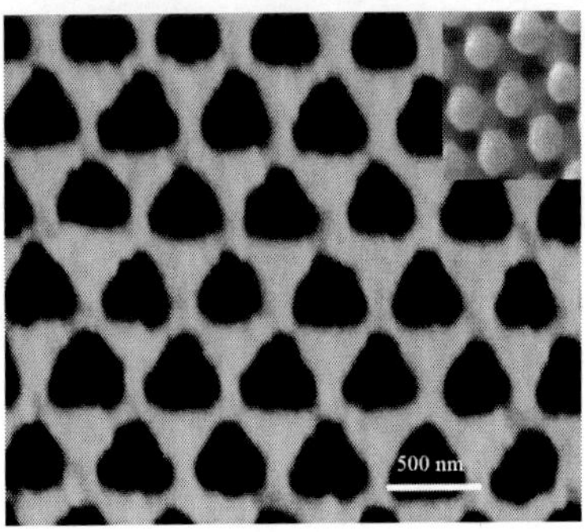

Vistas in Nanofabrication
Edited by Faiz Rahman

ISBN 978-981-4364-56-0 (Hardcover), 978-981-4364-57-7 (eBook)
www.panstanford.com

This SEM image shows the arrays of triangle silicon nanopores with high aspect ratio and tunable size by a combination of modified nanosphere lithography and deep reactive ion etching (DRIE). Nanospheres with a diameter of 500 nm were used in the experiment. A self-assembled bilayer nanosphere template is created as an etch mask for DRIE process. The inset shows that the size of nanopores is tuned by reducing the diameter of the top layer nanospheres using oxygen etching. The top spheres shrink much faster than the bottom spheres since the bottom spheres are covered by top spheres. Nanopore arrays with mean in-plane widths ranging from 120.9 nm to 343 nm and depths up to 2 μm have been achieved. In comparison with other nanofabrication techniques such as electron beam lithography and focused ion beam, which afford high resolution but lead to a low throughput, the etching technique is a cost-effective and high-throughput method for fabrication of arrays of high-aspect-ratio, size-tunable triangle nanopores. The nanopore size can not only be controlled by using different nanosphere size but also be tuned by the oxygen etching time. This approach offers excellent control over the nanopore size, aspect ratio, and pitch of the fabricated nanostructures. The approach can also be extended to fabricate nanopores of other semiconductor materials (e.g., germanium and gallium arsenide), glass, and so on, which may find more important applications in sensing, separations, cell growth, photonics, and labs-on-a-chip.

References

1. Chen X., Wei X., Jiang K. (2009) *Nanotechnology*, **20**, 425605 (5 pp).
2. Whitney A.V., Myers B.D., van Duyne R.P. (2004) *Nano Lett.*, **4,** 1507.

Self-Organizing Array of Spiral Nanoribbons

Li Zhang and Bradley J. Nelson
Institute of Robotics and Intelligent Systems, ETH Zurich, CH-8092, Switzerland
lizhang@ethz.ch, bnelson@ethz.ch

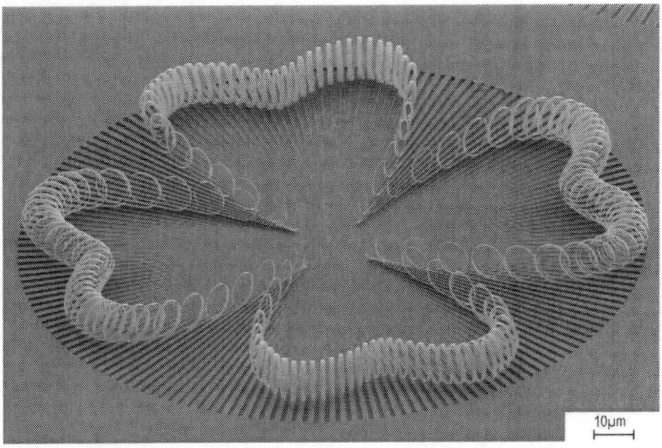

This micrograph shows a self-organizing array of spiral nanoribbons formed from a Si/Cr bilayer by the self-scrolling technique [1]. The process begins with the deposition of a p-type Si and Cr layer on an undoped Si substrate with a thickness of 35 nm and 10 nm, respectively. Then the Si/Cr bilayer is patterned into nanoribbons resembling the shape of a wagon wheel [2] by electron beam lithography. After selective etching of the underlying undoped Si substrate, the freestanding spirals form to release tensile strain in the Cr layer. The cloverleaf shape reflects the anisotropy of the selective etching of the Si (001) substrate. This self-scrolling technique can be used to fabricate a variety of 3-D micro- and nanostructures from various materials such as semiconductors, metals, and polymers [3]. Through this technique, the geometrical parameters of helical micro and nanostructures can be precisely controlled such as diameter, chirality, helicity angle, and pitch. Potential applications of these freestanding helical nanoribbons can be foreseen in many areas of nanotechnology, in particular MEMS/NEMS devices, flexible electronic devices, linear-to-rotary motion converters, wirelessly controlled micromanipulators, and swimming microrobots for biomedical applications [4].

References

1. Zhang L., Ruh E., Grutzmacher D., Dong L.X., Bell, D.J., Nelson B.J., Schoenenberger C. (2006) *Nano Lett.*, **6**, 1131–1317.
2. Seidel H., Csepregi L., Heuberger A., Baumgartel H. (1990) *J. Electrochem. Soc.*, **137**, 3612–3626.
3. (a) Prinz V.Y., Seleznev V.A., Gutakovsky A.K., Chehovskiy A.V., Preobrazhenskii V., Putyato M.A., Gavrilova T.A. (2000) *Physica E*, **6**, 828–831. (b) Schmidt O.G., Eberl K. (2001) *Nature*, **410**, 168.
4. (a) Zhang L., Abbott J.J., Dong L.X., Peyer K.E., Kratochvil B.E., Zhang H.X., Bergeles C., Nelson B.J. (2009) *Nano Lett.*, **9**, 3663–3667. (b) Dong L.X., Zhang L., Kratochvil B.E., Shou K., Nelson, B.J. (2009) *J. Microelectromech. Syst.*, **18**, 1047–1053. (c) Kratochvil B.E., Dong L.X., Zhang L., Nelson B.J. (2010) *J. Microscopy*, **237**, 122–135.

ZnO Nanowire Photonic Arrays

Jingbiao Cui
Department of Physics and Astronomy, University of Arkansas at Little Rock, Little Rock, Arkansas 72205, USA
jxcui@ualr.edu

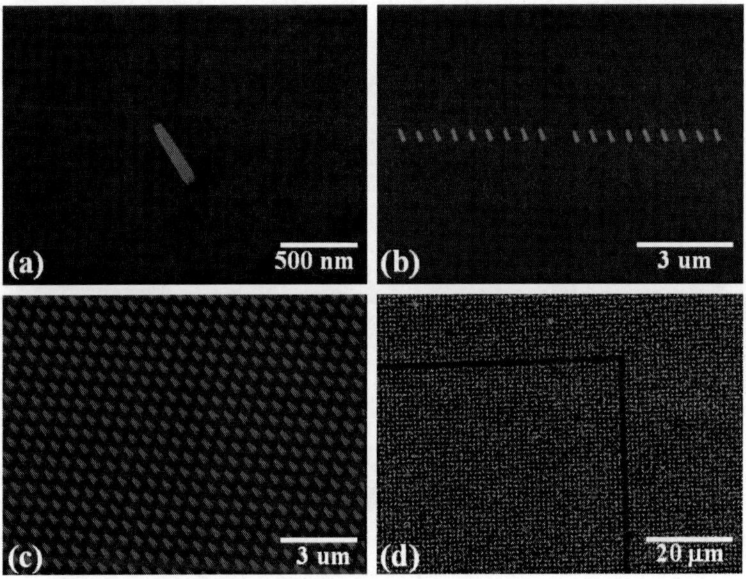

Photonic crystals constitute highly ordered and spatially aligned objects in one, two, or three dimensions, exhibiting a photonic bandgap within which a set of wavelengths are forbidden to propagate in the crystal, similar to the electronic bandgap in semiconductors [1]. They are potentially useful for photon control and manipulation in areas for inhibiting spontaneous emission, enhancing the performance of semiconductor lasers, providing a new means for guiding light, and playing a fundamental role in the integration and miniaturization of optical components.

The figure above shows ZnO nanowire photonic arrays fabricated by a combination of electrochemical growth and e-beam lithography [2]. Photonic nanostructures with desirable defects have been achieved with this innovative approach. The low growth temperature and mild conditions in the aqueous solution ensure the survival of polymethyl methacrylate (PMMA) template created by e-beam lithography. As shown in this figure, the growth of ZnO nanowires is precisely controlled in the desired locations to form zero, one, and two dimensional ordered arrays.

A single nanowire shown in Fig. (a) has length of about 1 μm and diameter of 100 nm. Figure (b) represents a row of ZnO nanowires with one missing nanowire in the middle of the row, which could potentially find applications as templates for construction of plasmonic resonators. Both experimental measurements and theoretical calculations have shown that the two-dimensional arrays of Fig. (c) have a photonic bandgap in the visible range [2]. Figure (d) is a two-dimensional ordered array of ZnO nanowires in which line defects were created along half of a row and a column. This 90° bending channel can be used as a waveguide for light within the two-dimensional photonic crystal for applications in large scale all-optical circuits [3].

References

1. Joannopoulos J.D., Meade R.D., Winn J.N. (1995). *Photonic Crystals* (Princeton University Press, New York).
2. Cui J.B., Gibson, U.J. (2007) *Nanotechnology*, **18**, 155302.
3. Lin S.Y., Chow E., Hietala V., Villeneuve P.R., Joannopoulos J.D. (1998) *Science*, 282, 274–276.

Nanostructure Cu_2O Films: Preparation, Characterization, and Properties

Ying Yu
Institute of Nanoscience and Nanotechnology, Central China Normal University, Luoyu Road 152, Hongshan District, Wuhan 430079, P. R. China
yuying@phy.ccnu.edu.cn

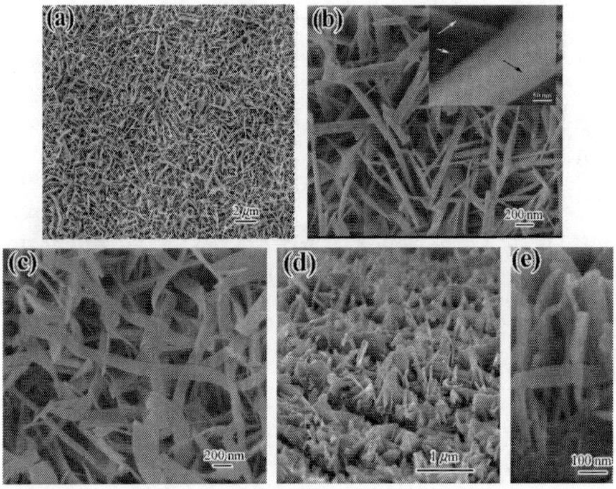

SEM images of Cu_2O aligned 2D film: (a–c) front images, (d and e) cross-sectional images. Reproduced with permission from Ma L. L. *et al.*, *J. Phys. Chem. C*, published by ACS 2008.

Cuprous oxide (Cu_2O) films on Cu foils with a two-dimensional (2-D) nanosheet structure were synthesized by polyol method with the assistance of acetamide and copper acetate as the copper source. The prepared Cu_2O nanosheets which are aligned on the Cu foils are single crystals with the thickness of ~20 nm. The addition of acetamide is crucial for the controllable synthesis of 2-D nanosheet structured Cu_2O films. Due to the aligned 2-D structure, the properties of the optical absorption and the charge carrier transport for the novel Cu_2O films are better than that for Cu_2O films with granular morphology. So, the films with 2-D aligned nanosheets morphology outperform the granular morphology in achieving higher photo-electrocurrent efficiency under simulated solar light, which makes them more suitable for solar energy applications. Additionally, besides polyol method, electrochemical and chemical deposition methods can be used for the preparation of Cu_2O nanostructure on substrates as well, and the obtained novel films show better electrochemical property and higher photocatalytic activity than normal bulk or granular films. These facile and efficient methods are proposed here towards the preparation of films with novel structure on different substrates for which a number of promising applications in various fields can be envisioned.

References

1. Ma L., Yan W., Li J., Wang E., Qiu M., Yu Y. (2008) *J. Phys. Chem. C.*, **112,** 18916–18922.
2. Yan L., Wang Y., Xiong L., Li J., Yip H., Wong P., Yu Y. (2009) *Chin. J. Inorg. Chem.*, **25**, 1960–1964.
3. Xiong L., Yang G., Yu H., Qiu M., Chen J., Yu Y. A simple approach for the synthesis of Cu_2O films with periodic morphology transformation onto different substrates (2010) *Thin Solid Film*, 518, 6738–6745.
4. Aref M.A., Xiong L., Abdulkarem A.M., Yan N., Yu Y.(2011) *Mater. Chem. Phys.* 127, 433–439.

Chemical Fabrication of Nanogaps for Single-Molecule Electronics

Titoo Jain, Kasper Nørgaard, and Thomas Bjørnholm
Nano-Science Center and Department of Chemistry, University of Copenhagen, Universitetsparken 5, DK-2100 Copenhagen Ø, Denmark
tb@nano.ku.dk

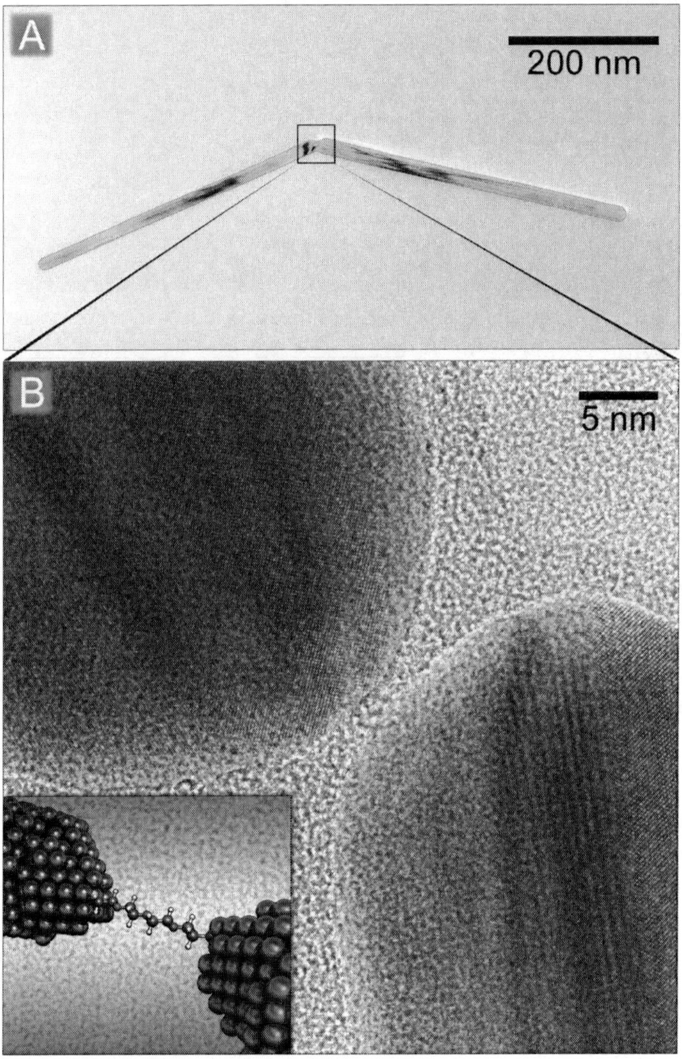

Single-molecule electronics is a field that seeks to exploit the possibility to integrate single molecules as building blocks in electronic circuits [1]. In order to investigate transport mechanisms through a single molecule it must be connected via electrodes to the macroscopic world. This goal can be achieved using so-called "bottom-up" approaches that can provide an elegant way to complement modern lithographic "top-down" fabrication techniques.

In our research, we focus on the chemical fabrication of nanogaps, where the molecule is already incorporated. We have previously shown how networks of gold nanoparticle wires can be created at the air–water interface [2, 3]. The next step is to exploit chemically prepared gaps between dimers of single crystalline gold nanorods (AuNRs) as their dimensions (500 nm × 20 nm) provide a feasible link between the easily addressable micrometer scale and the molecular length-scale [4]. Figure A shows a coloured transmission electron microscopy (TEM) image of a dimer of two AuNRs synthesized by linking spherical gold nanoparticle seeds with a dithiol-functionalized molecule. The seed particles are subjected to chemical conditions favoring growth resulting in the dimeric structures. The high-resolution TEM image in Fig. B reveals the atomic planes of the AuNRs together with the presence of a 1–2 nm nanogap perfectly suited for molecular electronics, while the inset shows a conceptual artistic drawing of the molecule incorporated in the gap of the two AuNRs.

References

1. Moth-Poulsen K., Bjørnholm T. (2009) *Nat. Nanotechol.*, **4**, 551–556.
2. Hassenkam T, Nørgaard K., Iversen L., Kiely C.J., Brust M., Bjørnholm T. (2002) *Adv. Mater.*, **14**, 1126–1130.
3. Hassenkam T., Moth-Poulsen K., Stuhr-Hansen N., Nørgaard K., Kabir M.S., Bjørnholm T. (2004) *Nano Lett.*, **4**, 19–22.
4. Jain T., Westerlund F., Johnson E., Moth-Poulsen K., Bjørnholm T. (2009) *ACS Nano*, **3**, 828–834.

Noncatalytic Synthesis of Carbon Nanotubes on SiC

Goknur Cambaz,[a] **Gleb Yushin,**[b] **and Yury Gogotsi**[c,*]

[a] *Department of Materials Science and Engineering, Cankaya University, Ankara, Turkey*

[b] *School of Materials Science and Engineering, Georgia Institute of Technology, Atlanta, GA 30332, USA*

[c] *Department of Materials Science and Engineering and A.J. Drexel Nanotechnology Institute, Drexel University, Philadelphia, PA 19104, USA*

*gogotsi@drexel.edu

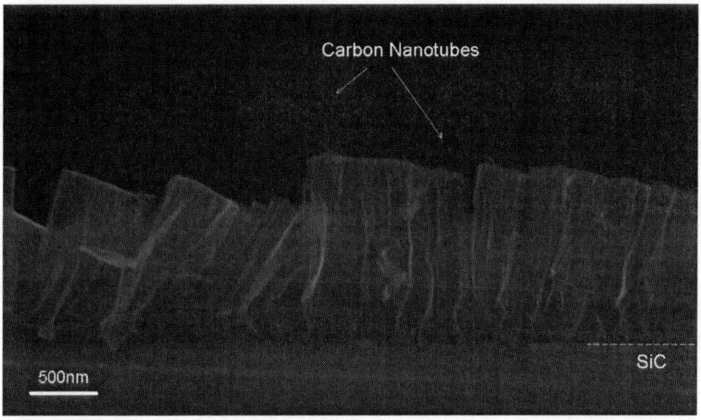

Control over the structure of materials on nanoscale can open numerous opportunities for the development of materials with controlled properties. Carbon, which is one of the most promising materials for nanotechnology, can be produced by many different methods. One of the most versatile, in terms of a variety of structures demonstrated (graphite, fullerene, porous amorphous carbon, nanotubes, grapheme, and diamond), is selective etching of SiC and other carbides [1]. Since the Si atoms are extracted layer by layer, atomic level control of the carbon structures can potentially be achieved without changing the size and shape of the sample. Carbon produced by this method is called Carbide-Derived Carbon (CDC).

Vertical CNT formation on both Si and C faces of single crystalline SiC wafers was achieved by vacuum annealing [2,3]. Carbon nanotubes (CNTs) synthesized on the Si face were characterized in detail (see figure). Strongly pronounced RBM modes in Raman spectra and Transmission electron microscope images showed that they are small-diameter nanotubes with 1–4 graphene walls. These nanotube arrays have a very high density and are catalyst-free with no internal closures. They show higher oxidation resistance and better mechanical properties [4] than CNT carpets deposited by catalytic chemical vapor deposition.

References

1. Yushin G., Gogotsi Y., Nikitin A., (2006) In *Nanomaterials Handbook* (ed. Y. Gogotsi) CRC Press. 237–280.
2. Kusunoki M., Rokkaku M., Suzuki T. (1997) *Appl. Phys. Lett.*, **71,** 2620–2622.

3. Cambaz G., Yushin G., Osswald S., Mochalin V., Gogotsi Y. (2008) *Carbon*, **46**, 841–849.

4. Pathak S., Cambaz Z.G., Kalidindi S.R., Swadener J.G., Gogotsi Y. (2009) *Carbon*, **47** 1969–1976.

Magnetic Microtags for Color MRI

Gary Zabow
National Institute of Standards & Technology (NIST), Boulder, Colorado, USA
National Institutes of Health (NIH), Bethesda, Maryland, USA
zabow@boulder.nist.gov

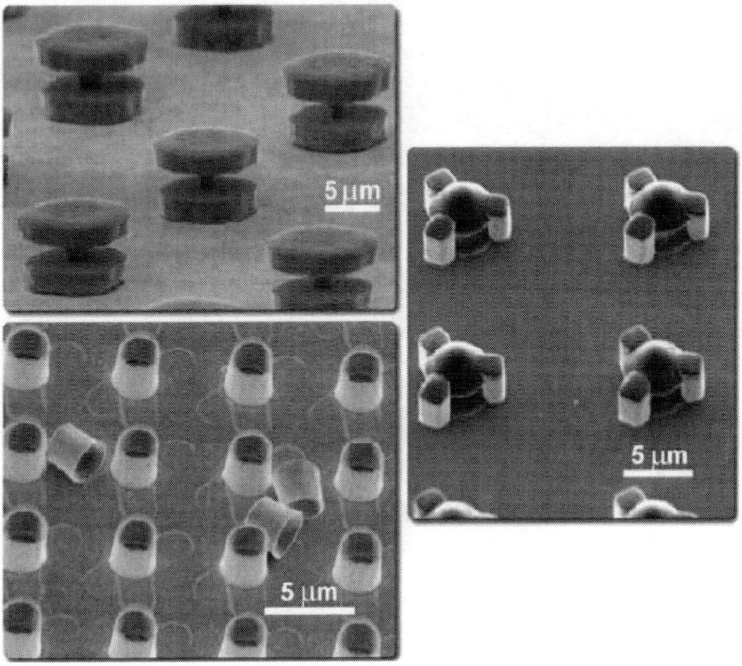

With its ability to see deep inside the body, magnetic resonance imaging (MRI) has become a widely used tool for medical diagnostics and imaging. To improve image quality, various contrast agents [1] are often used during MRI. These agents locally change the contrast, making a region of interest appear brighter or darker than the rest of the image. A new development in the field of contrast agents is specially designed magnetic micro- and nanotags that provide for changes in the color, not just in the contrast, of the image [2,3]. Unlike traditional chemically synthesized agents, the nanotags are fabricated by adapting microfabrication techniques [4] more commonly associated with the semiconductor industry. In this way, the nanotags can be made with special geometrical shapes that provide a unique radio-frequency signal, or effectively, a unique color, in an MRI. The included image shows scanning electron micrographs (SEM) of some of the geometries used for such magnetic micro- and nanotags. Clockwise from top are (i) double-disk structures consisting of two thick parallel plates of magnetizable material separated by a non-magnetic internal post; (ii) double-disk structures consisting of two thin layers of magnetizable material separated by external non-magnetic support posts; and (iii) magnetic shells in the shape of hollow cylinders. By varying the magnetic materials from which the structures are microfabricated, or by varying the structures' geometries, the nuclear magnetic resonance (NMR) frequency of surrounding water can be controllably shifted allowing the tags' effective colors to be arbitrarily tailored. Using series of such color-coded tags, it may become possible to use MRI to track different cell types as they migrate through the body or to monitor local physiological conditions inside the body.

References

1. Merbach A.E., Tóth E., eds. (2001) *The Chemistry of Contrast Agents in Medical Magnetic Resonance Imaging.* Wiley (UK).
2. Zabow G., Dodd S., Moreland J., Koretsky A. *Nature*, **453**, 1058–1063,

2008.

3. Zabow G., Dodd S.J., Moreland J, Koretsky A.P. (2009) *Nanotechnology*, **20**, 385301.
4. Madou M.J. (2002) *Fundamentals of Microfabrication: The Science of Miniaturization*. CRC Press (Florida).

Solution-Processible Fabrication of Gold Nanostructures

Xinping Zhang
Institute of Information Photonics Technology, Beijing University of Technology, Beijing 100124, P. R. China
College of Applied Sciences, Beijing University of Technology, Beijing 100124, P. R. China
zhangxinping@bjut.edu.cn

Solution-processible fabrication of gold nanostructures [1] using colloidal gold nanoparticles is a promising technique in the construction of plasmonic devices. In combination with interference lithography and annealing processes, one-dimensional gold nanowires (Figs. 1 and 2) [2] and two-dimensional gold nanocylinders (Fig. 3) or nanoholes (Fig. 4) can be fabricated with periodic arrangements. Using an annealing temperature in the range of 200–300°C, continuous gold nanowires can be obtained, which actually consist of tightly arranged gold nanoparticles, as shown in Fig. 1. When the annealing temperature is increased to above 450°C, the gold nanowires are broken into segments with an average length of a few micrometers, as shown in Fig. 2. However, this kind of high-temperature (>450°C) annealing process not only enables removal of the photoresist template but also enables formation of high-quality gold nanostructures, as can be seen in Figs. 2, 3, and 4, implying excellent optical response of particle plasmon resonance [3]. Periodically distributed gold nanostructures with each unit consisting of disorderly arranged gold nanoparticles can be obtained through controlling the concentration of the colloidal solutions and the size of the hole structures in photoresist master grating [4]. Direct annealing of the spin-coated film of colloidal gold nanoparticles can be defined as a new lithography-free technique for the fabrication of large-area gold nanostructures with excellent plasmonic response, as shown in Fig. 6.

Figure 1

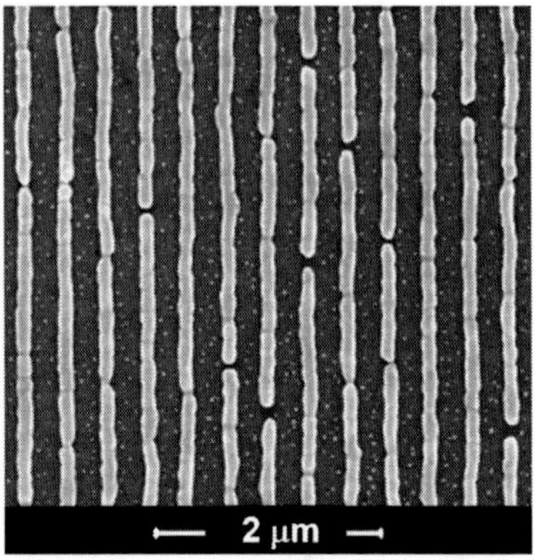

Figure 2

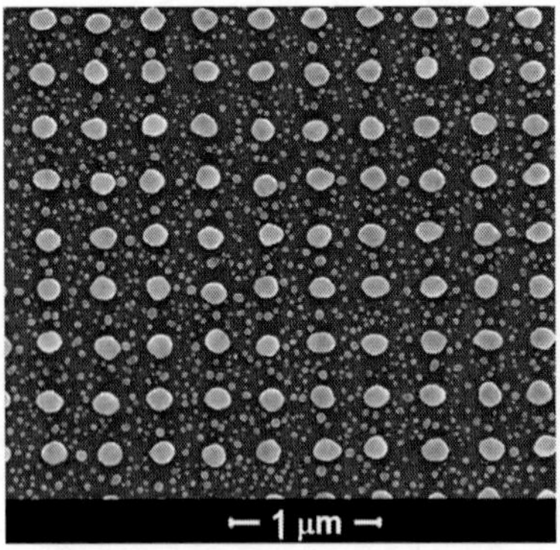

Figure 3

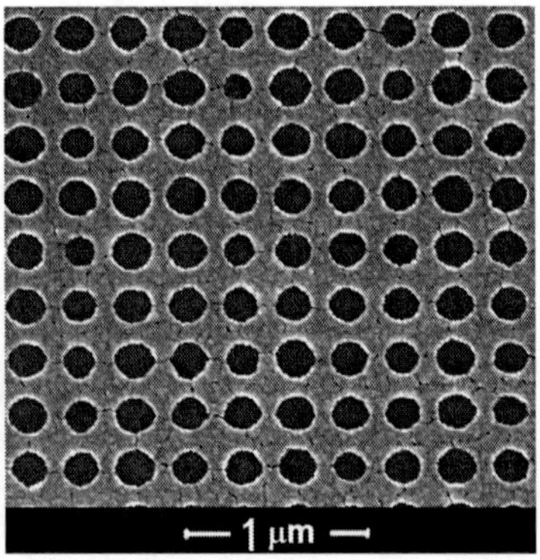

Figure 4

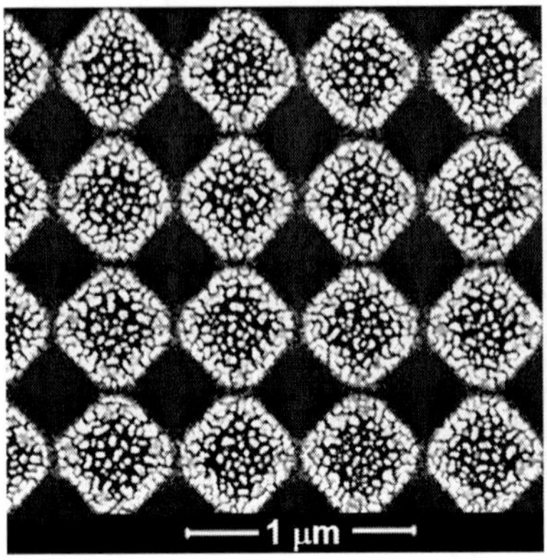

Figure 5

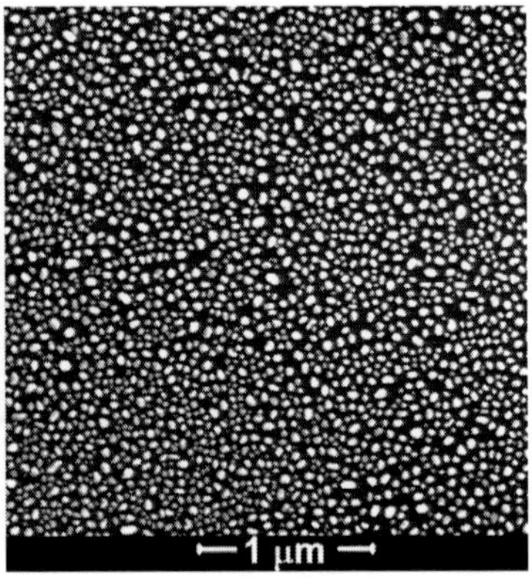

Figure 6

References

1. Zhang X., Sun B., Guo H., Nau D., Giessen H., Friend R.H. (2006) *Nano Lett.*, **6**, 651–655.
2. Zhang X., Liu H., Tian J., Song Y., Wang L., Song J., Zhang G. (2008) *Nanotechnology*, **19**, 285202 (1–7).
3. Zhang X., Liu H., Feng S. (2009) *Nanotechnology*, **20**, 425303 (1–8).
4. Zhang X., Sun B., Guo H., Tian J., Song Y., Wang L. (2007) *Chinese Opt. Lett.*, **5**, 657.

Resonant Tunneling Nanotransistor from III-V Compound Semiconductor Heterostructures

M. I. Lepsa,[a,*] J. Wensorra,[b] S. Trellenkamp,[a] J. Moers,[a] K. M. Indlekofer,[c] H. Lüth,[a] and D. Grützmacher[a]

[a] *Institute for Bio and Nanosystems(IBN-1) and JARA (Jülich-Aachen Research Alliance), Research Centre Jülich GmbH, D-52425 Jülich, Germany*

[b] *AREVA NP GmbH, Erlangen, Germany*

[c] *Information Technology and Electrical Engineering, FH Wiesbaden—University of Applied Sciences, Am Brückweg 26, D-65428 Rüsselsheim, Germany*

*m.lepsa@fz-juelich.de

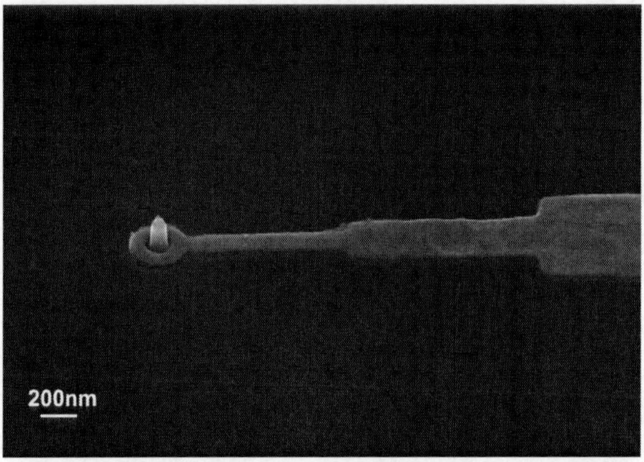

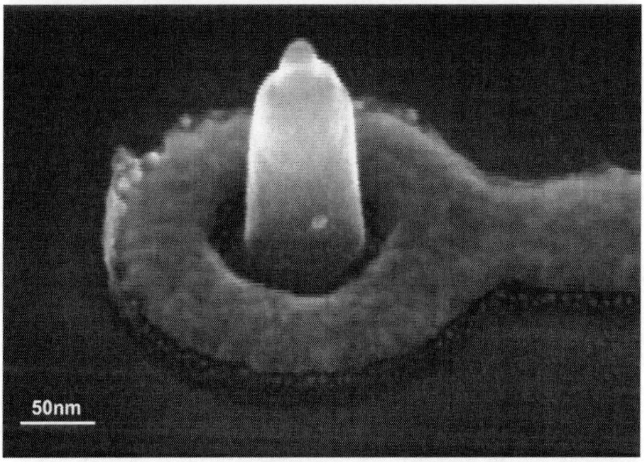

Modern semiconductor technology permits fabrication of structures and devices with specific dimensions in the deca-nanometer range. Structures based on quasi-1D nanocolumns are of major interest for basic research in the area of quantum transport and as prototype systems for novel opto-electronic and spintronic naonodevices. Nanocolumns can be processed either by employing self-organized growth modes or by lithographic patterning using a top-down approach. The later still offers the advantage of better realization of specific device patterns for large circuits and implementation of functional built-in heterostructures within the nanocolumns. An example in this sense refers to resonant tunneling transistors (RTTs). Fabrication of such devices starts with the processing of vertical sub-100 nm nanocolumns from an MBE grown GaAs/AlAs double barrier quantum well (DBQW) heterostructure [1]. The layer structure ends with a thin low temperature grown GaAs (LT-GaAs) layer which together with a Ti/Au metallization provides a very good nonalloyed shallow ohmic contact for nanodevices [2]. The top down approach is based on electron-beam (E-B) lithography of hydrogen silsesquioxan (HSQ), as mask material, and subsequent dry etching processes for the column definition. HSQ is used to planarize and isolate the device electrodes as well. The all-around metal gate is processed at the level of the DBQW structure by precise etching of the dielectric and E-B lithography. For the lithographic step, a special alignment procedure is used which gives a reproducible alignment accuracy of less than 10 nm. The gate is positioned at a distance of about 20 nm relative to the semiconductor nanocolumn.

The processed nano-RTT by the top down approach show good electrical performances with pronounced negative differential resistance in the I-V characteristics due to forwards collimation of the propagating electrons in a doping-induced potential saddle point in front of the DBQW structure [3]. In this context, a whole new field of room temperature nanoelectronics both for logic and memory applications is opened, where low power consumption is combined with high speed and less circuit complexity for high functionality.

References

1. Wensorra J., Lepsa M.I., Trellenkamp S., Moers J., Indlekofer K.M., Lüth H. (2009) *Nanotechnology*, **20**, 465402 (6 pp.).
2. Wensorra J., Lepsa M.I., Indlekofer K.M., Förster A., Jaschinsky P., Voigtländer B., Pirug G., Lüth H. (2006). Ohmic contacts for GaAs based nanocolumns, *Phys. Status Solidi A*, **203**, 3559–3564.
3. Wensorra J., Indlekofer K.M., Lepsa M.I., Förster A., Lüth H. (2005) *Nano Lett.*, **5**, 2470–2475.

Index

Color Insert

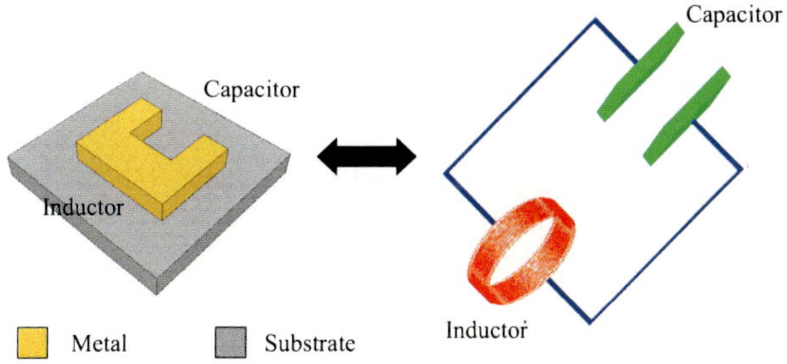

(a) Split ring resonator (SRR)

(b) Equivalent circuit

Figure 3.1

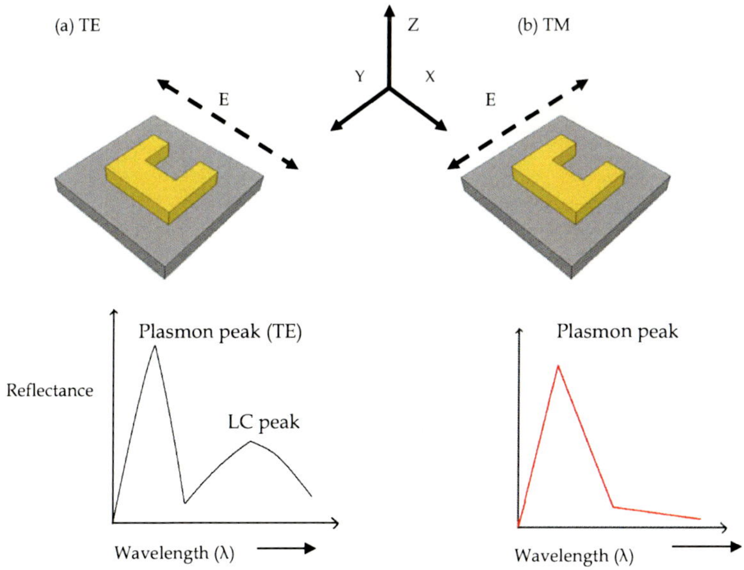

Figure 3.2

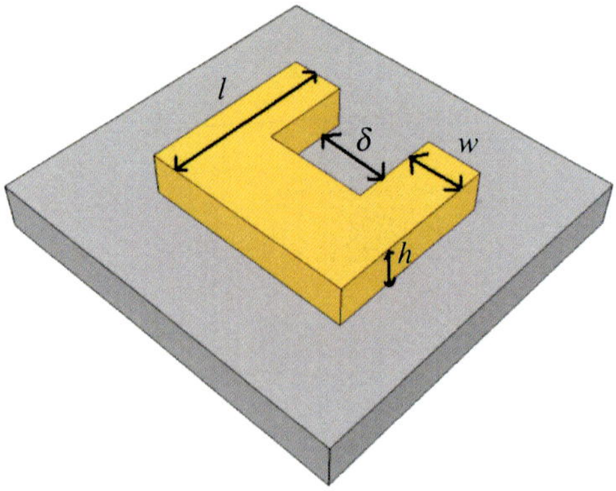

Figure 3.3

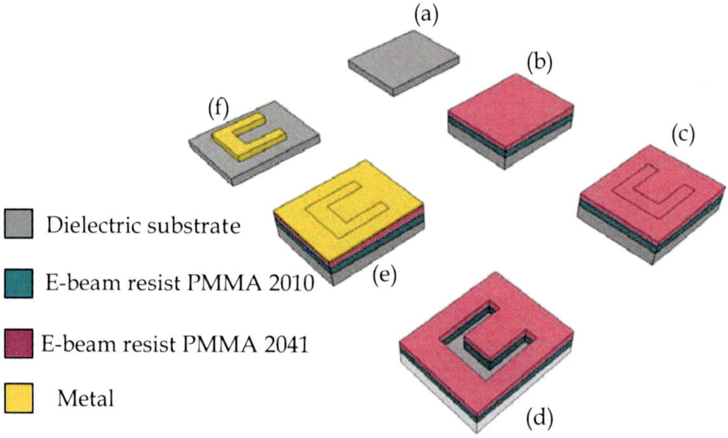

Figure 3.4

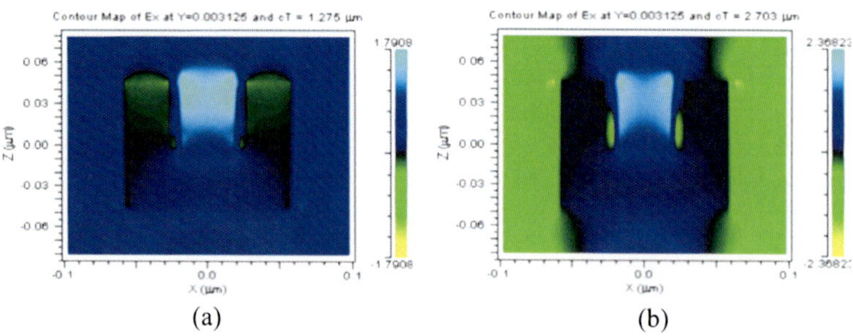

Figure 3.8

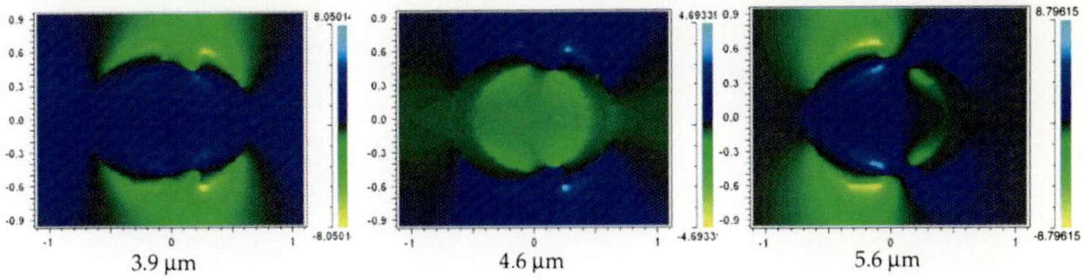

Figure 3.11

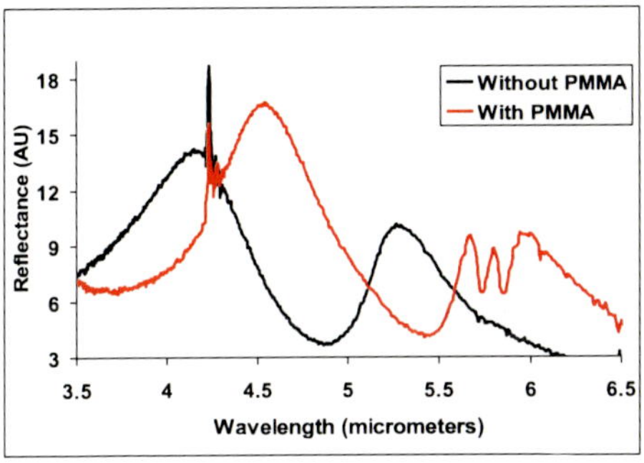

Figure 3.14

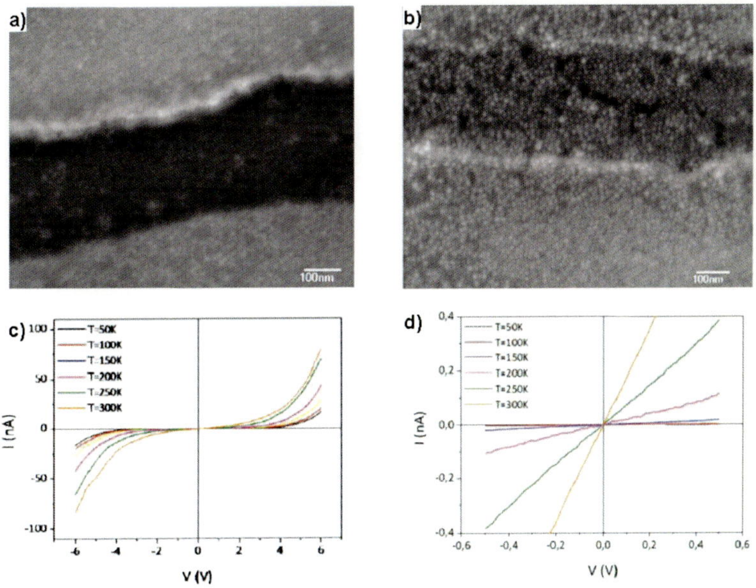

Figure 6.4

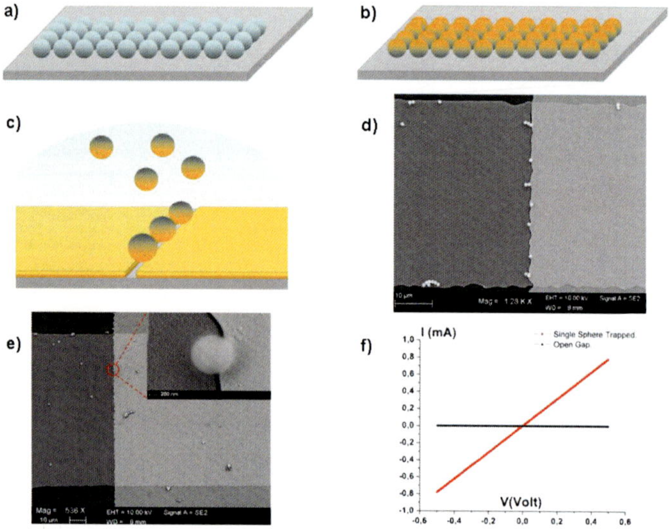

Figure 6.5

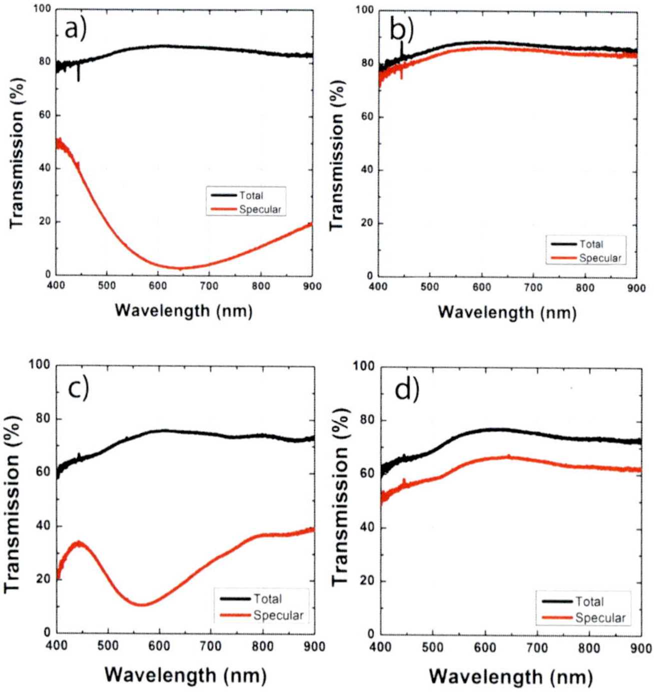

Figure 7.12

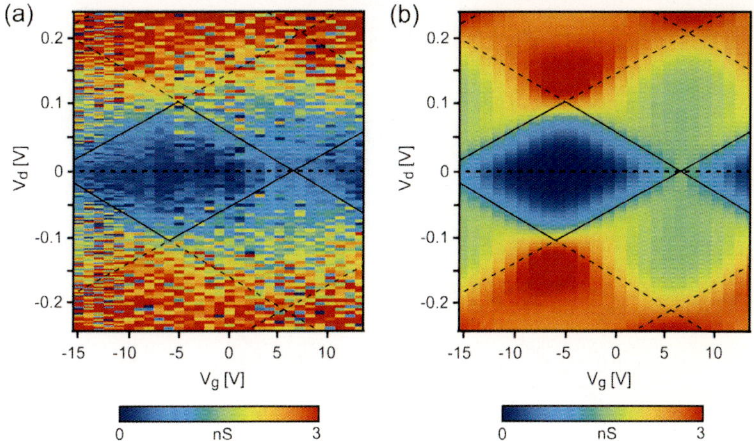

Figure 8.5

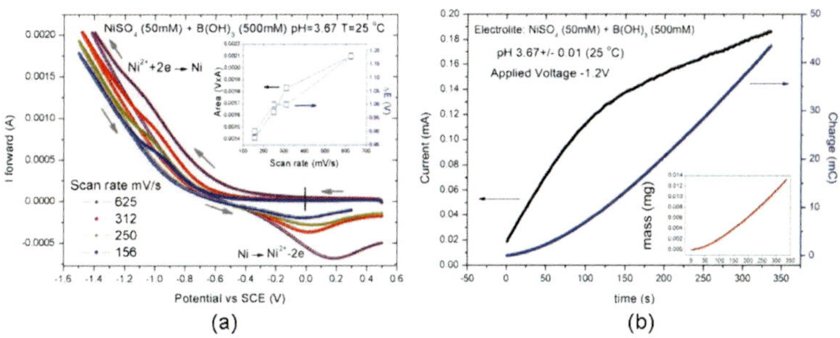

Figure 8.7

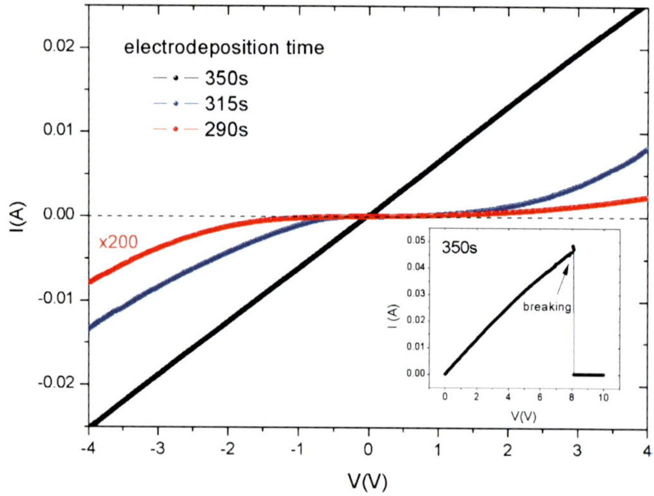

Figure 8.12

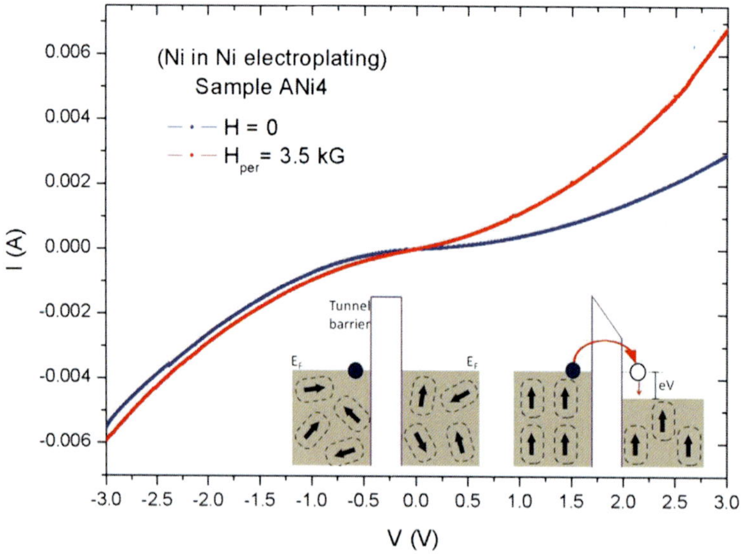

Figure 8.13

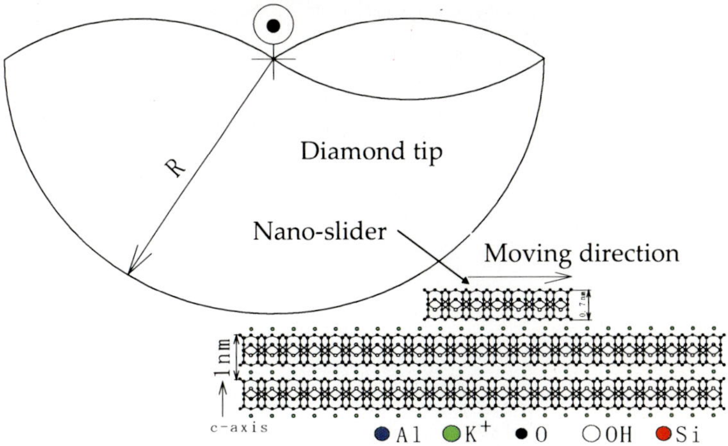

(a) Model of nano-slider

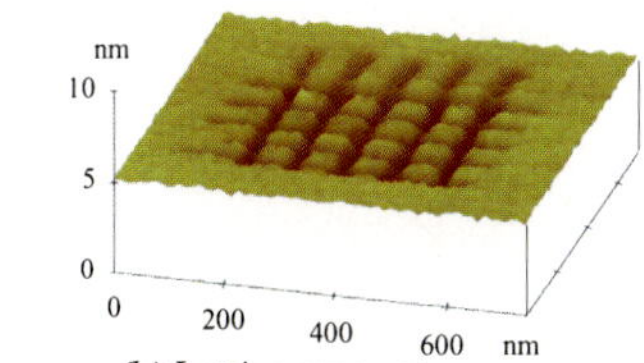

(b) Lattice grooves

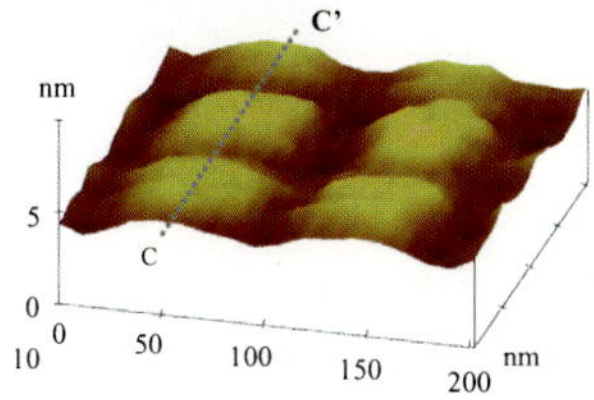

(c) Before tip action

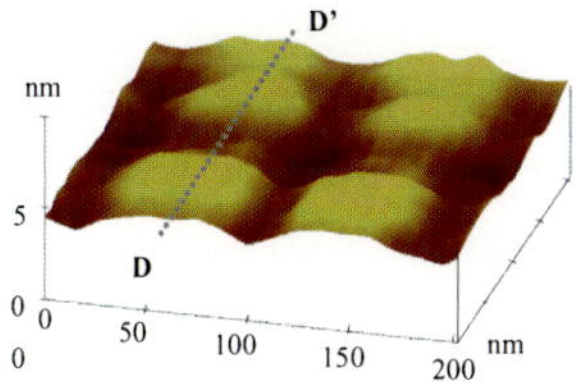

(d) After tip action

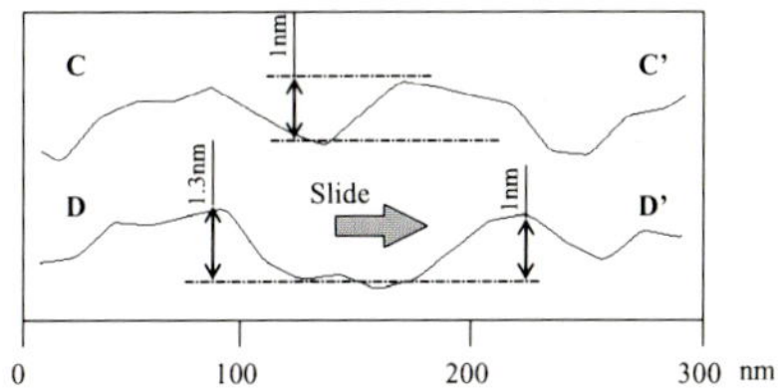

(e) Cross section of C –C′ and D –D′

Figure 9.11

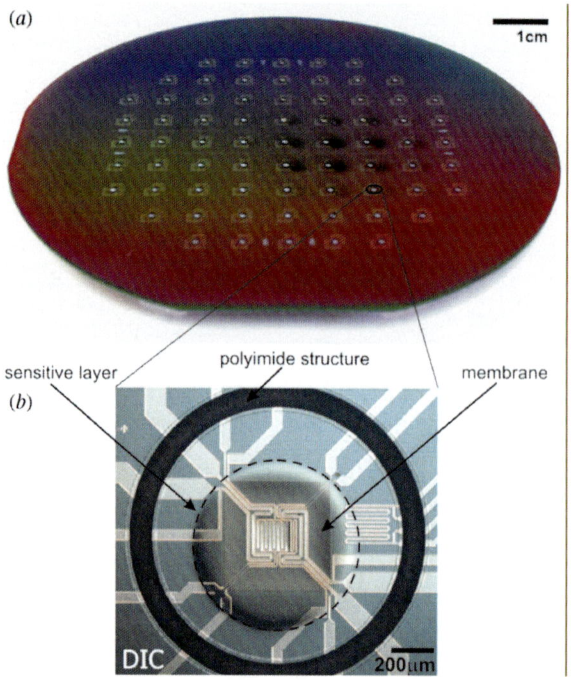

Figure 10.9

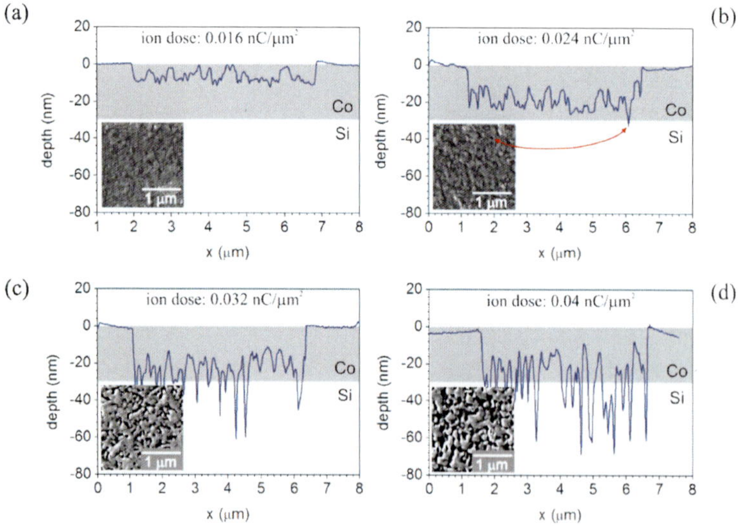

Figure 11.3